工业和信息化
精品系列教材 · 大数据技术

Hadoop
平台搭建与应用

第 2 版 | 微课版

米洪 张鸰 / 主编
杨琼 郑莹 王宇博 / 副主编

Hadoop Platform Construction
and Application

人民邮电出版社
北京

图书在版编目（CIP）数据

Hadoop平台搭建与应用：微课版 / 米洪, 张鸰主编. -- 2版. -- 北京：人民邮电出版社, 2024.8
工业和信息化精品系列教材. 大数据技术
ISBN 978-7-115-64077-2

Ⅰ. ①H… Ⅱ. ①米… ②张… Ⅲ. ①数据处理软件－高等学校－教材 Ⅳ. ①TP274

中国国家版本馆CIP数据核字(2024)第063653号

内 容 提 要

本书基于企业级大数据处理应用项目，采用"任务驱动式"的编写方式，指导读者基于 Hadoop 生态系统进行大数据平台的搭建与应用。全书共包括 8 个项目，分别是认识大数据、Hive 环境搭建与基本操作、ZooKeeper 环境搭建与应用、HBase 环境搭建与基本操作、Hadoop 生态组件的安装与使用、Hadoop HA 集群搭建、Ambari 搭建与集群管理，以及 Hadoop 平台应用综合案例。

本书实用性和可操作性较强，讲解精练，通俗易懂，可作为高等院校大数据技术及其他相关专业的教材，也可作为从事大数据分析等工作的技术人员的参考书。

◆ 主　编　米　洪　张　鸰
　　副主编　杨　琼　郑　莹　王宇博
　　责任编辑　顾梦宇
　　责任印制　王　郁　焦志炜

◆ 人民邮电出版社出版发行　北京市丰台区成寿寺路 11 号
　　邮编　100164　电子邮件　315@ptpress.com.cn
　　网址　https://www.ptpress.com.cn
　　三河市君旺印务有限公司印刷

◆ 开本：787×1092　1/16
　　印张：15　　　　　　　　　　　　　　2024 年 8 月第 2 版
　　字数：365 千字　　　　　　　　　　　2024 年 8 月河北第 1 次印刷

定价：59.80 元

读者服务热线：(010)81055256　印装质量热线：(010)81055316
反盗版热线：(010)81055315
广告经营许可证：京东市监广登字 20170147 号

前 言

1. 缘起

党的二十大报告提出，必须坚持科技是第一生产力、人才是第一资源、创新是第一动力，深入实施科教兴国战略、人才强国战略、创新驱动发展战略，开辟发展新领域新赛道，不断塑造发展新动能新优势。当前，大数据正对全球生产、流通、分配、消费活动及经济运行机制和社会生活方式产生重大影响，我们要应用大数据相关技术从海量数据中发现新知识、创造新价值。传统的数据处理方式和处理平台已经无法满足大数据应用发展的需求，以云计算技术为支撑的大数据处理平台的出现，为落实大数据应用提供了可行的思路和方案。Hadoop 是一个能够让用户轻松搭建和使用的分布式计算平台，用户可以轻松地在 Hadoop 上分析和处理海量数据。

2. 使用

（1）教学内容课时安排：本书建议授课 64 课时，课时安排如下表所示。

课时安排

项目序号	项目名称	项目任务	模块课时	项目课时
1	认识大数据	任务 1.1 认知大数据，完成系统环境搭建	2	10
		任务 1.2 Docker 安装及常用命令	4	
		任务 1.3 Hadoop 环境搭建	4	
2	Hive 环境搭建与基本操作	任务 2.1 Hive 的安装与配置	4	6
		任务 2.2 Hive 的应用	2	
3	ZooKeeper 环境搭建与应用	任务 3.1 ZooKeeper 的安装与配置	4	6
		任务 3.2 ZooKeeper CLI 操作	2	
4	HBase 环境搭建与基本操作	任务 4.1 HBase 的安装与配置	4	6
		任务 4.2 HBase Shell 操作	2	
5	Hadoop 生态组件的安装与使用	任务 5.1 Sqoop 的安装与使用	4	16
		任务 5.2 Pig 的安装与使用	4	
		任务 5.3 Flume 的安装与使用	4	
		任务 5.4 Kafka 的安装与使用	2	
		任务 5.5 Flink 的安装与使用	2	

续表

项目序号	项目名称	项目任务		模块课时	项目课时
6	Hadoop HA 集群搭建	任务 6.1	Hadoop HA 集群环境搭建	4	6
		任务 6.2	Hadoop HA 集群的启动与自动故障转移测试	2	
7	Ambari 搭建与集群管理	任务 7.1	搭建 Ambari 平台	4	6
		任务 7.2	使用 Ambari 管理 Hadoop 集群	2	
8	Hadoop 平台应用综合案例	任务 8.1	本地数据集上传到 Hive 中	2	8
		任务 8.2	使用 Hive 进行简单的数据分析	2	
		任务 8.3	Hive、MySQL、HBase 数据的互导	2	
		任务 8.4	流数据处理的简单应用	2	
合计					64

（2）课程资源获取：本书是云计算技术应用、大数据技术、人工智能技术应用等相关专业的产教融合系列教材之一，数字化课程资源丰富。对于与本书配套的运行脚本、电子教案等资源，读者可以登录人邮教育社区（www.ryjiaoyu.com）下载。

3．致谢

本书由南京交通职业技术学院的米洪、张鸽任主编并完成统稿工作，由浙江工业职业技术学院的杨琼和南京交通职业技术学院的郑莹、王宇博任副主编。在编写本书的过程中，编者参阅了国内外同行编写的相关著作和文献，谨向各位作者致以深深的谢意！

由于编者水平有限，书中难免存在欠妥之处，希望广大读者批评、指正。编者联系方式：njcimh@njitt.edu.cn。

编　者

2024 年 2 月

目 录

项目 1 认识大数据 ………………………… 1

学习目标 ………………………………… 1

项目描述 ………………………………… 1

任务 1.1 认知大数据，完成系统环境
搭建 ………………………………… 1

  任务描述 …………………………… 1

  任务目标 …………………………… 2

  知识准备 …………………………… 2

  任务实施 …………………………… 16

任务 1.2 Docker 安装及常用命令 …… 24

  任务描述 …………………………… 24

  任务目标 …………………………… 24

  知识准备 …………………………… 24

  任务实施 …………………………… 26

任务 1.3 Hadoop 环境搭建 …………… 33

  任务描述 …………………………… 33

  任务目标 …………………………… 33

  知识准备 …………………………… 33

  任务实施 …………………………… 34

项目小结 ………………………………… 48

课后练习 ………………………………… 48

项目 2 Hive 环境搭建与基本操作 …… 49

学习目标 ………………………………… 49

项目描述 ………………………………… 49

任务 2.1 Hive 的安装与配置 ………… 49

  任务描述 …………………………… 49

  任务目标 …………………………… 49

  知识准备 …………………………… 50

  任务实施 …………………………… 53

任务 2.2 Hive 的应用 ………………… 58

  任务描述 …………………………… 58

  任务目标 …………………………… 59

  知识准备 …………………………… 59

  任务实施 …………………………… 66

项目小结 ………………………………… 74

课后练习 ………………………………… 75

项目 3 ZooKeeper 环境搭建与应用 … 76

学习目标 ………………………………… 76

项目描述 ………………………………… 76

任务 3.1 ZooKeeper 的安装与配置 …… 76

  任务描述 …………………………… 76

任务目标 ································· 76

　　知识准备 ································· 77

　　任务实施 ································· 82

任务 3.2　ZooKeeper CLI 操作 ············ 94

　　任务描述 ································· 94

　　任务目标 ································· 94

　　知识准备 ································· 94

　　任务实施 ································· 96

项目小结 ····································· 100

课后练习 ····································· 101

项目 4　HBase 环境搭建与基本

　　　　操作 ································· 102

学习目标 ····································· 102

项目描述 ····································· 102

任务 4.1　HBase 的安装与配置 ·········· 102

　　任务描述 ································· 102

　　任务目标 ································· 102

　　知识准备 ································· 103

　　任务实施 ································· 109

任务 4.2　HBase Shell 操作 ················ 119

　　任务描述 ································· 119

　　任务目标 ································· 119

　　知识准备 ································· 119

　　任务实施 ································· 120

项目小结 ····································· 126

课后练习 ····································· 127

项目 5　Hadoop 生态组件的安装与

　　　　使用 ································· 128

学习目标 ····································· 128

项目描述 ····································· 128

任务 5.1　Sqoop 的安装与使用 ·········· 128

　　任务描述 ································· 128

　　任务目标 ································· 128

　　知识准备 ································· 129

　　任务实施 ································· 132

任务 5.2　Pig 的安装与使用 ··············· 136

　　任务描述 ································· 136

　　任务目标 ································· 136

　　知识准备 ································· 136

　　任务实施 ································· 141

任务 5.3　Flume 的安装与使用 ·········· 149

　　任务描述 ································· 149

　　任务目标 ································· 149

　　知识准备 ································· 149

　　任务实施 ································· 150

任务 5.4　Kafka 的安装与使用 ·········· 152

　　任务描述 ································· 152

　　任务目标 ································· 152

　　知识准备 ································· 152

　　任务实施 ································· 155

任务 5.5　Flink 的安装与使用 ············ 157

　　任务描述 ································· 157

任务目标 ········· 157

知识准备 ········· 157

任务实施 ········· 160

项目小结 ············· 162

课后练习 ············· 162

项目 6　Hadoop HA 集群搭建 ········· 164

学习目标 ············· 164

项目描述 ············· 164

任务 6.1　Hadoop HA 集群环境
　　　　　搭建 ········· 164

任务描述 ········· 164

任务目标 ········· 164

知识准备 ········· 165

任务实施 ········· 166

任务 6.2　Hadoop HA 集群的启动与
　　　　　自动故障转移测试 ········· 178

任务描述 ········· 178

任务目标 ········· 178

知识准备 ········· 179

任务实施 ········· 179

项目小结 ············· 182

课后练习 ············· 182

项目 7　Ambari 搭建与集群管理 ········· 183

学习目标 ············· 183

项目描述 ············· 183

任务 7.1　搭建 Ambari 平台 ········· 183

任务描述 ········· 183

任务目标 ········· 183

知识准备 ········· 184

任务实施 ········· 186

任务 7.2　使用 Ambari 管理 Hadoop
　　　　　集群 ········· 207

任务描述 ········· 207

任务目标 ········· 207

知识准备 ········· 207

任务实施 ········· 207

项目小结 ············· 213

课后练习 ············· 213

项目 8　Hadoop 平台应用综合案例 ···· 214

学习目标 ············· 214

项目描述 ············· 214

任务 8.1　本地数据集上传到
　　　　　Hive 中 ········· 214

任务描述 ········· 214

任务目标 ········· 214

任务实施 ········· 215

任务 8.2　使用 Hive 进行简单的数据
　　　　　分析 ········· 218

任务描述 ········· 218

任务目标 ········· 218

任务实施 ········· 218

任务 8.3　Hive、MySQL、HBase 数据的互导 …… 220

　　任务描述 …… 220

　　任务目标 …… 220

　　任务实施 …… 220

任务 8.4　流数据处理的简单应用 …… 225

　　任务描述 …… 225

　　任务目标 …… 225

　　任务实施 …… 226

项目小结 …… 231

课后练习 …… 232

项目 ① 认识大数据

学习目标

【知识目标】
识记大数据的概念和特征。
领会大数据处理与分析流程。

【素质目标】
具有严谨细致的工作态度和工作作风。
具有良好的团队协作意识和业务沟通能力。

【技能目标】
熟悉大数据分析与处理工具。
学会 CentOS 的安装和常用命令。
学会 Docker 的安装和常用命令。
学会 Hadoop 的安装与配置。

项目描述

大数据正在催生以数据资产为核心的多种商业模式,产生了巨大的应用价值。数据的生成、分析、存储、分享、检索、消费构成了大数据的生态系统,其中每一个环节都产生了不同的需求,新的需求又驱动技术创新和方法创新。大数据技术正在融合社会应用,使数据参与决策,发挥大数据真正有效的价值,进而影响人们未来的生活模式。近年来,伴随着物联网的兴起、移动应用的流行和社交媒体的快速发展,大数据技术展现出其独有的时代特性,广泛应用于用户群体细分、数据搜索、个性推荐、用户关系管理等方面。大数据技术的巨大的延伸价值使其成为时代焦点,引起了人们的关注。

大数据技术是收集、整理、处理大容量数据集,并从中获得所需的非传统战略和技术的总称。由于 Hadoop 已经成为应用最广泛的大数据框架技术之一,因此大数据的相关技术主要围绕 Hadoop 展开,涵盖 Hadoop、MapReduce、HDFS 和 HBase 等技术。

本项目引导学生完成 CentOS 的安装、Docker 的安装和 Hadoop 在 3 种运行模式下的安装与配置,并学习 CentOS 和 Docker 的常用命令。

任务 1.1 认知大数据,完成系统环境搭建

任务描述

(1)学习大数据的相关知识,熟悉大数据的定义、大数据的基本特征及大数据处理与分析的相关技术、工具或产品等。

(2)完成系统环境搭建,为 Hadoop 搭建做好准备工作。

 **任务目标**

（1）熟悉大数据的概念和特征。
（2）熟悉大数据分析流程和工具的使用。
（3）学会 CentOS 的安装。
（4）学会 CentOS 常用命令的使用。
（5）学会 CentOS 的相关配置。

知识准备

1. 大数据背景知识

大数据是信息技术（Information Technology，IT）领域备受瞩目的名词，在全球引领了新一轮数据技术革命的浪潮，通过 2012 年的蓄势待发，2013 年被称为"世界大数据元年"，标志着世界正式步入了大数据时代。移动互联网、物联网及传统互联网等每天都会产生海量的数据，人们可以使用适当的统计分析方法对收集来的海量数据进行分析，将它们加以汇总和理解，以求最大化地开发数据的功能，发挥数据的作用。

想要系统地认知大数据，必须要全面而细致地分解它，接下来将从 3 个层面展开介绍大数据，如图 1-1 所示。

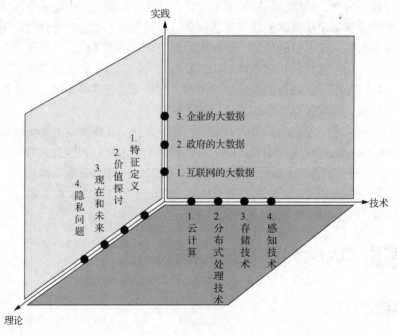

图 1-1 大数据的 3 个层面

第 1 个层面是理论。理论是认知的必经途径，也是被广泛认同和传播的基线。人们可以从大数据的特征定义出发，去理解行业对大数据的整体描绘和定性；从对大数据的价值探讨出发,去深入解析大数据的价值所在;从大数据的现在和未来出发,去洞悉大数据的发展趋势;从大数据的隐私问题出发,去审视人和数据之间的长久博弈。

第 2 个层面是技术。技术是大数据的价值体现手段和前进的基石。人们可以从云计算、分布式处理技术、存储技术和感知技术出发,理解大数据从采集、处理、存储到形成结果的整个过程。

第 3 个层面是实践,实践是大数据的最终价值体现。人们可以从互联网的大数据、政府的大数据和企业的大数据 3 个方面出发,描绘大数据已经展现的美好景象及即将实现的蓝图。

(1)从理论层面认知大数据

① 大数据的特征定义。最早提出大数据时代到来的是麦肯锡咨询公司,它是美国首屈一指的咨询公司,是研究大数据的先驱。其在报告"Big data: The next frontier for innovation, competition, and productivity"中给出的大数据的定义如下:大数据指的是数据规模超出常规的数据库工具获取、存储、管理和分析能力的数据集。

业界(IBM 最早定义)将大数据的特征归纳为 4 个"V"。

- 数据体量巨大,即 Volume。大数据的起始计量单位至少是 PB(2^{10}TB)、EB(2^{20}TB)或 ZB(2^{30}TB)。
- 数据类型繁多,即 Variety。例如,网络日志、视频、图片、地理位置等都是不同数据类型的信息。
- 价值密度低,商业价值高,即 Value。由于数据采集的不及时、数据样本的不全面、数据不连续等,可能会导致数据失真,但当数据量达到一定规模时,可以通过更多的数据实现更真实、全面的反馈。
- 处理速度快,即 Velocity。对大数据的处理速度要求较高,一般要在秒级时间范围内给出分析结果,否则时间太长,数据就失去了价值。这个速度要求是大数据处理技术和传统的数据挖掘技术最大的区别之一。

本书认同大数据研究机构高德纳(Gartner)对大数据的定义——大数据是指无法在一定时间范围内用常规软件工具进行获取、管理和处理的数据集,是需要使用新处理模式才能处理的具有更强决策力、洞察发现力和流程优化能力的海量、高增长率和多样化的信息资产。

② 大数据的价值探讨。大数据已经成为当今社会中不可忽视的资源和工具,其在经济领域中具有巨大的潜力和价值。具体来说,大数据的价值主要体现在以下方面。首先,在大数据的支持下,企业可以获取更全面、准确的用户信息,从而能够更精准地了解用户的需求和购买习惯;其次,大数据可以帮助企业预测市场趋势,为企业提供更精准的市场营销和销售策略,提高销售额和市场占有率,提供更高效的运营管理策略;最后,大数据的有效利用能够帮助企业及时发现和应对市场变化,缓解竞争压力,降低商业风险和损失等。

在科学研究领域,大数据被广泛应用于各个学科的研究中,为科学家提供了更广阔的研究空间和更深入的认识,大数据的应用可以更全面、真实地反映事物的本质和变化规律,帮助科学家发现更多的规律和模式,提供更准确的实验结果,搭建更准确的预测模型,加快学术研究的进展和成果的应用,推动科学的发展和创新。

大数据还具有重要的社会价值,通过对大数据的应用,政府可以更准确地了解社会需求和问题，制定更有效的政策和措施，提高公共资源的利用效率，实现城市管理的智能化，促进社会治理和社会公正，提供更精细化、智能化的公共服务。

综上，大数据具有广泛而深远的价值。它不仅为企业带来了经济效益和商业机会，也为科研人员提供了更广阔的研究领域和更深入的认识，同时还为社会治理和公共服务提供了更科学有效的手段。随着大数据技术和应用的不断发展，相信大数据的价值将会更加凸显对大数据的有效利用将为社会创造更多的福祉。

③ 大数据的现在和未来。现在，大数据的应用价值已在各行各业凸显。大数据能够帮助政府实现市场经济调控、公共卫生与安全防范、灾难预警等工作；大数据能够帮助城市预防犯罪、实现智慧交通、提升应急能力；大数据能够帮助医疗机构建立患者的疾病风险跟踪机制、帮助医药企业提升药品的临床使用效果；大数据能够帮助航空公司节省运营成本、帮助电信企业提升售后服务质量、帮助保险企业识别欺诈骗保行为、帮助快递公司监测和分析运输车辆的故障以便提前预警维修、帮助电力公司有效识别和预警即将发生故障的设备等。

不管大数据的核心价值是不是"预测"，基于大数据形成决策的模式已经为不少的企业和机构带来了盈利和声誉。

未来，在大数据领域最具有价值的将是以下两种。

- 拥有大数据思维的人。这类人可以将大数据的潜在价值转换为实际利益。
- 还没有被大数据触及的业务领域。这些领域是还未被挖掘的"金矿"，即所谓的"蓝海"。

④ 大数据的隐私问题。大数据时代，隐私保护是人们必须面对的问题。当人们在不同的网站上填写个人信息后，这些信息可能已经扩散出去了；当人们莫名其妙地遭受各种邮件、电话、短信的骚扰时，可能是因为自己的电话号码、邮箱、生日、购买记录、收入水平、家庭住址等私人信息，早就被各种商业机构非法存储或转卖给其他有需要的企业或个人了；当微博、微信、QQ这些社交平台掌握着数亿用户的各种信息时，用户很难保护好个人隐私，即使用户在某个地方删除了相关信息，但这些信息也许早就已经被其他用户转载或保存，甚至有可能已经被平台保存为快照，并提供给其他用户进行搜索了。因此，在大数据的背景下，很多人在积极地抵制无底线的数字化，这种大数据和个体之间的博弈还会一直继续下去。

个人隐私的保护需要政府、企业和个人共同努力。可以从以下方面加强对个人隐私的保护。

- 加强法律法规的制定和执行。为了保护个人隐私，政府应加强对大数据的监管，制定相应的法律法规，并加强执法力度。
- 加强数据安全保护。为了防止个人信息泄露，各个机构和企业应加强数据安全保护措施，包括加密存储、访问控制、数据备份等。同时，对用户个人数据的收集和使用应明确告知用户，并取得用户的同意。
- 推动隐私保护技术的研发和应用。随着大数据时代的到来，隐私保护技术也在不断发展。例如，利用差分隐私技术可以在保护个人隐私的同时，保持数据的可用性和准确性。政府和企业应加大对隐私保护技术的研发和应用力度，确保个人数据安全。
- 加强教育和提升意识。在大数据时代，个人隐私保护需要广泛的社会参与。政府和

项目1　认识大数据

媒体应加强对个人隐私保护意识的教育及宣传工作，提高公众对隐私问题的关注和重视。同时，个人也应增强自我保护意识，避免在网络和社交媒体上过度暴露个人信息。

大数据时代的隐私问题是一个复杂而严峻的挑战。只有通过多方合作和共同努力，才能在大数据时代实现对个人隐私的有效保护。

（2）从技术层面认知大数据

① 云计算。大数据常和云计算联系在一起，因为实时的大型数据集分析需要分布式处理框架向数十、数百甚至数万台计算机分配工作，大数据和云计算的结合能够实现对海量数据的挖掘。如今，在谷歌、亚马逊等一批互联网企业的引领下，一种行之有效的模式被创建出来，即云计算提供基础架构平台，大数据应用运行在这个平台之上。

行业内普遍认为大数据和云计算的关系如下：没有大数据的信息积淀，云计算的计算能力再强大，也难以找到用武之地；没有云计算的计算能力，大数据的信息积淀再丰富，终究也只是镜花水月。

云计算和大数据之间的关系如图 1-2 所示。

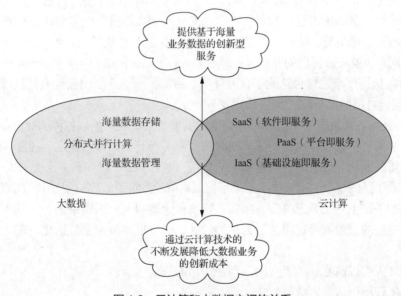

图 1-2　云计算和大数据之间的关系

大数据和云计算的结合可以提供更多基于海量业务数据的创新型服务，并通过云计算技术的不断发展降低大数据业务的创新成本。云计算与大数据最明显的区别表现在以下两个方面。

- 两者在概念上有所不同。云计算改变了 IT，大数据改变了业务。然而，大数据必须有云计算作为基础架构，才能顺畅运行。
- 两者的目标受众不同。云计算是首席信息官等关心的技术层，是一种进阶的 IT 解决方案。大数据是首席执行官关注的业务层产品，大数据的决策者是业务层。

② 分布式处理技术。分布式处理技术可以将位于不同地点、具有不同功能、拥有不同数据的多台计算机用通信网络连接起来，在控制系统的统一管理下，协调其共同完成信息处理任务。

大数据分布式处理技术的典型代表是 Hadoop，它有一个 MapReduce 软件框架，能以一种可靠、高效、可伸缩的方式对大数据进行分布式处理。MapReduce 是谷歌提出的一种云计算核心计算模式，是一种分布式计算技术，也是简化的分布式编程模型。MapReduce 模式的主要思想是将要执行的数据（如程序）自动分割，将复杂的计算任务拆解成 Map（映射）和 Reduce（规约）两个主要步骤。在数据被分割后，通过 Map 函数将数据映射成不同的区块，分配给计算机集群处理，以进行分布式计算，再通过 Reduce 函数将结果汇总，从而输出开发者需要的结果。

③ 存储技术。大数据技术可以抽象地分为大数据存储和大数据分析，大数据存储的目的是支撑大数据分析。到目前为止，大数据存储和大数据分析属于不同的计算机技术领域。大数据存储致力于研发可以扩展至 PB 甚至 EB 级别的数据存储平台，而大数据分析则关注如何在最短时间内处理大量不同类型的数据集。

大数据存储服务商的典型代表是亚马逊。亚马逊的 S3 是一种面向互联网的存储服务，此服务旨在让开发人员更轻松地进行网络规模计算。它提供了一个简明的 Web 服务界面，用户可以在任何地点访问 Web 服务界面，存储和检索任意大小的数据。此服务让所有开发人员都能访问同一个具备高扩展性、高可靠性、高安全性和高性价比的基础设施，亚马逊的网站都借助此基础设施来运行。亚马逊的 S3 云的跨地域存储对象数量已达到万亿级别，而且性能表现出色，同时，亚马逊提供的专业云计算服务——亚马逊网络服务（Amazon Web Service，AWS）的对象执行请求数量也达到了百万峰值。目前，全球范围内已经有超过 10 万家企业通过 AWS 运行自己的部分或者全部日常业务。

④ 感知技术。大数据的采集和感知技术的发展是紧密联系的。各种工业设备、汽车、电表上有大量的数码传感器，这些传感器随时测量和传递有关位移、速度、温度、湿度乃至空气中化学物质的变化等信息，并产生海量的数据。

随着智能手机的普及，感知技术迎来了发展高峰期，除了地理位置感知方法被广泛应用，一些新的感知手段也开始登上舞台，例如，指纹传感器和人脸识别系统等。其实，这些感知信息被逐渐捕获的过程就是世界被数据化的过程，一旦世界被完全数据化，那么世界的本质就是信息了。

所以说"人类以前延续的是文明，现在传承的是信息"。

（3）从实践层面认知大数据

① 互联网的大数据。互联网是大数据发展的前沿阵地，随着互联网的发展，人们似乎都习惯了将自己的生活通过网络进行数据化，以方便分享、记录和回忆。

互联网的大数据类型很难清晰地界定，先看看中国三大互联网公司——百度、阿里巴巴、腾讯拥有的大数据类型。

百度拥有两种类型的大数据：用户搜索行为特征的需求数据、爬虫和百度阿拉丁平台获取的公共网络数据。百度通过对网页数据进行爬取，对网页内容进行组织和解析，并对这些网络信息进行语义分析，产生对搜索需求的精准理解，以便从海量数据中得到所需要的结果，这个过程实质上就是一个数据的获取、组织、分析和挖掘的过程。

阿里巴巴拥有交易数据和信用数据，在这两种类型的数据中更容易挖掘出商业价值。除此之外，阿里巴巴还通过投资等方式掌握了部分社交数据、位置数据，如微博社交数据和高德地图相关数据等。

腾讯拥有用户关系数据和基于此产生的社交数据。通过分析这些数据可以了解用户的生活和行为，还可能挖掘出政治、社会、文化、商业、健康等领域的信息。

在国外，除了行业知名的谷歌之外，也涌现了很多专门经营大数据类型产品的公司，这里主要介绍以下几家公司。

- Tableau。它致力于将海量数据以可视化的方式展现出来。Tableau 为数字媒体提供了新的数据展示方式，为用户提供了免费工具，使用户在没有编程知识背景的情况下也能制造出数据专用图表。它还能对数据进行分析，并提供有价值的建议。
- ParAccel。它向美国执法机构提供数据分析服务，例如，它对 15000 个有犯罪前科的人进行跟踪，并向执法机构提供参考价值较高的犯罪预测结果，被称为"犯罪的预言者"。
- Qlik。Qlik 旗下的 QlikView 是一种商业智能领域的自主服务工具，能够应用于科学研究和艺术等领域。为了帮助开发者对这些领域的数据进行分析，Qlik 还提供了对原始数据进行可视化处理等功能的工具 Qlik Sense。
- GoodData。GoodData 希望帮助用户从数据中挖掘财富，它主要面向商业用户和 IT 企业高管，为其提供数据存储、性能报告、数据分析等工具。

下面简要归纳一下互联网中大数据的典型代表。

- 用户行为数据。用于精准广告投放、内容推荐、行为习惯和喜好分析、产品优化等业务。
- 用户消费数据。用于精准营销、信用记录分析、活动促销、理财等业务。
- 用户地理位置数据。用于线下推广、商家推荐、交友推荐等业务。
- 互联网金融数据。用于点对点网络借款、支付、信用、供应链金融等业务。
- 用户社交数据。用于潮流趋势分析、流行元素分析、受欢迎程度分析、社会问题分析等业务。

② 政府的大数据。在我国，政府各个部门都掌握着构成社会基础的原始数据，如气象数据、金融数据、信用数据、电力数据、煤气数据、自来水数据、道路交通数据、客运数据、刑事案件数据等。这些数据在每个政府部门中看起来都是单一的、静态的。但是，如果政府将这些数据关联起来，并对这些数据进行有效地关联分析和统一管理，那么这些数据一定可以产生无法估量的价值。

具体来说，现在城市都在走向智能化，如智能电网、智慧交通、智慧医疗、智慧环保，这些都依托于大数据，可以说大数据是城市智能化的核心。大数据分析为智慧城市的各个领域提供决策支持。在城市规划方面，通过对城市地理、气象等自然信息和经济、社会、文化、人口等人文社会信息的挖掘，大数据分析可以为城市规划提供决策，强化城市管理服务的科学性和前瞻性。在交通管理方面，通过对道路交通信息的实时挖掘，大数据分析能有效缓解交通拥堵，并快速响应突发状况，为城市交通的良性运转提供科学的决策依据。在安防与防灾方面，通过对大数据的挖掘，可以及时发现人为或自然灾害、恐怖事件，提高应急处理能力和安全防范能力。

③ 企业的大数据。随着互联网的发展和 IT 技术的更新换代，大数据的应用平台越来越多，种类也越来越丰富，大数据正在悄无声息地改变着企业的运营模式、市场的导向。

国内的大数据厂商，如华为、腾讯、阿里巴巴和百度，在大数据分析技术领域分别研制了许多功能丰富且成熟的产品。

- 华为。
  - 计算领域产品。

华为坚定不移地投入计算产业，并在以下方面取得了一定成绩。第一，架构创新。华为投资基础研究，推出达芬奇架构，用创新的处理器架构来匹配算力的增速。第二，华为投资全场景处理器族，包括面向通用计算的鲲鹏系列、面向人工智能（Artificial Intelligence，AI）计算的昇腾系列、面向智能终端的麒麟系列和面向智慧屏的鸿鹄系列等。第三，华为坚持"有所为，有所不为"的商业策略。不直接对外销售处理器，而是以云服务面向用户，以部件为主面向合作伙伴，优先支持合作伙伴发展整机。第四，华为积极构建开放生态。

鲲鹏通用计算平台提供基于鲲鹏处理器的 TaiShan 服务器、鲲鹏主板及开发套件，适配各行业厂商的多样性计算、绿色计算需求，致力于打造最强算力平台。硬件厂商可以基于鲲鹏主板发展自有品牌的产品和解决方案；软件厂商可以基于 openEuler 开源操作系统（Operating System，OS）以及配套的数据库、中间件等平台软件发展其应用软件和服务；鲲鹏开发套件可帮助开发者加速应用迁移和算力升级。

基于华为昇腾（HUAWEI Ascend）系列 AI 处理器的华为 Atlas 人工智能计算解决方案通过模块、板卡、小站、服务器、集群等丰富的产品形态，打造面向"端、边、云"的全场景 AI 基础设施方案，覆盖深度学习领域的推理和训练全流程。

  - OceanStor 数据存储。

数字经济时代，数据作为新型的生产要素，已经成为基础性和战略性资源，数据应用的蓬勃发展，需要有强大、安全、可靠的数据存储设施保驾护航。在不确定性、复杂性、多元性激增的时代，华为 OceanStor 坚持用秉承"以数据为中心，构建多样化数据应用可靠存储底座"的理念，应需而变的海纳能力，来融合和识别难以被洞察的数据需求，释放数据智慧，用基于融合广泛、灵活应变、绿色环保、面向未来的能力，来助力用户从容应对数字时代的多重挑战，实现利于当下、着眼未来的永续发展。

- 腾讯。

腾讯云大数据技术已经从以离线计算为代表的第一代、以实时计算为代表的第二代、以机器学习为代表的第三代，发展到如今以隐私计算、数智融合以及云原生为代表的第四代。

腾讯云在首届大数据峰会上公布，其大数据平台算力规模已经突破千万核，日实时计算量达百万亿级，日运行容器数超亿级，日计算数据量数百 PB，服务的企业用户数超 2 万家，开源社区代码贡献量超 800 万行，进一步展示了其在大数据领域的强大实力。

腾讯云原副总裁黄世飞表示，腾讯云基于全新的技术架构、数据治理理念以及产品能力，从底层的大数据基础引擎、中层的一站式大数据开发治理平台，再到上层的智能推荐、隐私计算和商务智能（Business Intelligence，BI）应用，已经构筑起国内领先的大数据产品矩阵。

  - 深化云原生能力，大数据基础产品"开箱即用"。

腾讯云大数据基于开源、开放理念，以及内部单集群超万节点的大规模技术实践，沉淀了国内领先的企业级云原生大数据技术架构的构建能力。弹性 MapReduce（Elastic MapReduce，EMR）、Elasticsearch Service（ES）、云数据仓库（Cloud Data Warehouse，CDW）、数据湖计算（Data Lake Compute，DLC）、流计算 Oceanus、数据集成 DataInLong 等构成了腾讯云大数据基础产品的核心引擎。

- 基于全新治理理念,WeData 全面升级。

一站式大数据开发治理平台 WeData,基于业内独创的 DataOps 数据开发模式,通过协同、效率、一体三大核心优势,实现了数据集成、数据资产、数据开发、数据服务等全方位的数据治理能力。WeData 能帮助企业大幅提升数据管理效率和数据质量,为业务创造更多价值。

- 大数据和 AI 融合,实现数据与业务闭环。

腾讯云大数据通过和自主研发的人工智能平台 TI-ONE 紧密结合,让经过腾讯内部海量业务训练的智能化平台融入大数据业务中。新一代智能推荐平台采用业界领先的 AI 技术和推荐算法,能够为信息流分类页推荐、短视频个性化推荐、信息流"猜你喜欢"等场景提供强大技术支撑。

为打破数据孤岛,实现数据协同,腾讯云大数据团队自主研发 Angel PowerFL 隐私计算框架和基于 Spark + MQ 的隐私计算模式,为政务、金融、广告营销等众多行业提供了数据安全保障,并因此连续三年获得 iDASH 国际隐私计算大赛冠军。

腾讯云副总裁刘煜宏表示,在未来,腾讯云大数据还将在自适应计算架构、流批一体的实时湖仓、大数据"自动驾驶平台"、云原生大数据等方向上进一步发力,携手合作伙伴,为用户提供更智能、更安全的大数据产品服务。

- 阿里巴巴。
  - 云数据仓库 MaxCompute。

MaxCompute 是面向分析的企业级软件即服务(Software as a Service,SaaS)模式的云数据仓库,它以 Serverless 架构提供快速、全托管的在线数据库服务。MaxCompute 消除了传统数据平台在资源扩展性和弹性方面的限制,最小化用户运维投入,使用户可以以经济、高效的方式分析、处理海量数据。数以万计的企业正基于 MaxCompute 进行数据计算与分析,将数据高效转换为业务洞察。

- 实时数仓引擎 Hologres。

Hologres 是一站式实时数仓引擎,它支持海量数据的实时写入、实时更新、实时分析,支持标准 SQL(兼容 PostgreSQL 协议),支持 PB 级数据联机分析处理(Online Analytical Processing,OLAP)与即席查询(Ad Hoc Query),支持高并发、低延迟的在线数据服务。Hologres 与 MaxCompute、Flink、DataWorks 深度融合,提供离线/在线一体化全栈数仓解决方案。

- 大数据开发治理平台 DataWorks。

DataWorks 基于 MaxCompute、Hologres、EMR、企业数据云平台(Cloudera Data Platform,CDP)等大数据引擎,为数据库、数据湖、湖仓一体等解决方案提供统一的全链路大数据开发治理平台。作为阿里巴巴数据中台的建设者,DataWorks 从 2009 年起不断沉淀阿里巴巴大数据建设方法论,同时与数万名来自政务、金融、零售、互联网、能源、制造等领域的用户携手,助力产业数字化升级。

- 实时计算 Flink 版。

阿里云实时计算 Flink 版是阿里云基于 Apache Flink 构建的企业级、高性能实时大数据处理系统,由 Apache Flink 创始团队官方出品,拥有全球统一商业化品牌,完全兼容开源 Flink

应用程序接口（Application Program Interface，API），提供丰富的企业级增值功能。

■ 数据湖构建。

数据湖构建（Data Lake Formation，DLF）作为云原生数据湖架构的核心组成部分，可帮助用户快速地构建云原生数据湖架构。数据湖构建提供湖上元数据统一管理、企业级权限控制功能，可无缝对接多种计算引擎，打破数据孤岛，洞察业务价值。

■ 数据总线。

数据总线（DataHub）服务是阿里云提供的流式数据（Streaming Data）服务，它提供流式数据的发布（Publish）和订阅（Subscribe）功能，让企业可以轻松构建基于流式数据的分析和应用。

● 百度。

百度智能云大数据产品丰富，旨在为企业提供从构建新型数据基础设施、深度挖掘数据价值，到保障数据安全的全流程大数据解决方案，助力企业数字化创新升级。

■ 百度云原生湖仓架构。

百度云原生湖仓架构为企业解决大数据基础设施构建中的数据存储、计算、处理、治理开发等问题。

■ 百度智能云对象存储。

百度智能云对象存储（Baidu Object Storage，BOS）是一款稳定、安全、高效、高可拓展的云存储服务，支持标准、低频、冷和归档存储等多种存储类型，满足多场景的存储需求。用户可以将任意数量和形式的非结构化数据存入 BOS，并对数据进行管理和处理。

■ 百度托管大数据平台。

百度托管大数据平台（Baidu MapReduce，BMR）是全托管的 Hadoop/Spark 集群，可以按需部署和弹性扩展集群，企业只需专注于大数据处理、分析、报告，拥有多年大规模分布式计算技术积累的百度运维团队将全权负责集群运维。

■ 百度数据仓库 Palo。

百度数据仓库 Palo 是基于开源 Apache Doris 构建的企业级大规模并行处理（Massively Parallel Processing，MPP）云数据仓库，可有效地支持在线实时数据分析，具有简单易用、流批一体、高可用性等特征。

■ 基于 BES 的日志检索架构。

基于 BES 的日志检索架构将可被直接检索的数据存储周期由 7 天升级为最高 180 天，存储成本降低近 90%，其热数据检索速度实现了秒级响应。

■ 数据可视化 Sugar BI。

Sugar BI 是百度自助 BI 报表分析和制作可视化数据大屏的工具，可直连多种数据源，通过丰富的图表和拖曳式编辑帮助用户轻松生成可视化界面，并以大屏的方式呈现，让数据信息更直观。同时，Sugar BI 融合了百度语音、语义识别等多种 AI 技术，用户通过语音的方式就可以快速获取想要的数据。

■ 百度智能云全功能 AI 开发平台 BML。

百度智能云全功能 AI 开发平台 BML（Baidu Machine Learning）具备高效的算力管理和调度、高性能数据科学引擎、自动机器学习、丰富的建模方式四大核心功能，提供从数据源管理、数据标注、数据集存储、数据预处理、模型训练生产到模型管理、预测推理服务管理等全流程开发支持，让用户预测未来有据可依。

■ 百度隐私计算的核心产品"百度点石"。

"百度点石"是基于安全多方计算、联邦学习、机密计算、安全数字沙箱的隐私计算引擎。它打造出一套集数据安全治理规则与隐私计算工具于一体的解决方案，通过技术与规则的巧妙结合，在"数据可用不可见"与"数据不动算法动"的基础上，赋能企业机构合法合规地采集数据、存储数据、挖掘数据，实现在合理保护用户隐私基础上的数据驱动经济发展。

2. 大数据处理流程

具体的大数据处理方法有很多，根据长时间的实践，可以总结出一个基本的大数据处理流程。整个处理流程可以概括为 4 步，分别是采集、导入和预处理、统计与分析、挖掘。

（1）采集

大数据的采集是指利用多个数据库来接收客户端（如 Web、App 或者传感器等）的数据，并且用户可以通过这些数据库来进行简单的查询和处理工作。例如，电商行业会使用传统的关系数据库 MySQL 和 Oracle 等存储每一笔业务数据。除此之外，键-值（Key-Value）对型数据库（如 Redis）、文档型数据库（如 MongoDB）、图形数据库（如 Neo4j）等 NoSQL 数据库也常用于大数据的采集。

大数据采集过程的主要特点和挑战是并发量高，因为同时可能有成千上万的用户在进行访问和业务操作，例如，火车票售票网站 12306 和购物网站淘宝，它们的并发访问量在峰值时可达到上百万。因此需要在采集端部署大量数据库才能支撑采集操作，并且如何在这些数据库之间进行负载均衡和分片也是需要深入思考和设计的。

（2）导入和预处理

虽然采集端本身会有很多数据库，但是如果要对这些数据库中的海量数据进行有效地分析，则应该将这些来自前端的数据导入一个集中的大型分布式数据库或者分布式存储集群，并且在导入的基础上做一些简单的清洗和预处理工作。有一些用户会在导入时使用来自推特、领英等公司开源的流式计算系统 Apache Storm、分布式发布订阅消息系统 Kafka 等对数据进行流式计算，来满足部分业务的实时计算需求。

导入和预处理过程的主要特点和挑战是导入的数据量大，每秒的导入量经常会达到百兆比特，甚至千兆比特级别。

（3）统计与分析

统计与分析主要利用分布式数据库，或者分布式计算集群来对存储于其内的海量数据进行普通地分析和分类汇总等，以满足大多数常见的分析需求。在这方面，一些实时性需求会用到易安信（EMC，一家美国信息存储资讯科技公司）的分布式数据库 Greenplum、Oracle 的新一代数据库云服务器 Exadata 以及基于 MySQL 的列式存储数据库 Infobright 等，而一些批处理或者基于半结构化数据的需求可以使用 Hadoop。

统计与分析过程的主要特点和挑战是分析涉及的数据量大，其对系统资源，特别是输入/输出（Input/Output，I/O）会有极大的占用。

（4）挖掘

和统计与分析过程不同的是，挖掘一般没有预先设定好的主题，主要是在现有数据上进行基于各种算法的计算，从而起到预测（Predict）的效果，以便实现一些高级别数据分析的需求。在挖掘中比较典型的算法有用于聚类的 K-Means、用于统计学习的支持向量机（Support Vector Machine，SVM）和用于分类的朴素贝叶斯，使用的工具主要有 Hadoop 的 Mahout 等。

挖掘过程的主要特点和挑战是其使用的算法很复杂，计算涉及的数据量和计算量都很大，且常用数据挖掘算法以单线程为主。

数据来自各个方面，在面对庞大而复杂的大数据时，选择一个合适的处理工具显得尤为重要。"工欲善其事，必先利其器"，一个好的大数据分析处理工具不仅可以使工作事半功倍，还可以让人们在竞争日益激烈的大数据时代，挖掘大数据价值，及时调整战略方向。

3. 大数据分析工具

（1）Hadoop

Hadoop 是一个能够对大量数据进行处理的分布式计算平台，其以一种可靠、高效、可伸缩的方式进行数据处理。Hadoop 2.0 的架构如图 1-3 所示。

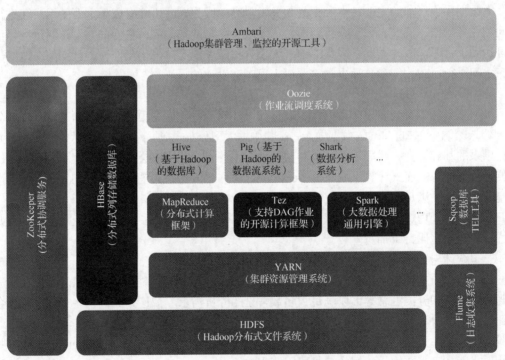

图 1-3　Hadoop 2.0 的架构

① HDFS（Hadoop 分布式文件系统）。HDFS 是 Hadoop 体系中数据存储和管理的基础。它是一个高度容错的系统，能检测和应对硬件故障，用于在低成本的通用硬件上运行。HDFS 简化了文件的一致性模型，通过流式数据访问，提供高吞吐量应用程序数据的访问功能，适

用于带有大型数据集的应用程序。

② YARN（集群资源管理系统）。YARN 是一种新的 Hadoop 资源管理器，它是一个通用资源管理系统，可为上层应用提供统一的资源管理和调度，它的引入为集群在利用率、资源统一管理和数据共享等方面带来了巨大好处。

③ MapReduce（分布式计算框架）。MapReduce 是一种计算模型，用于进行大数据的计算。其中，Map 对数据集上的独立元素进行指定的操作，生成 Key-Value 形式的中间结果；Reduce 则对中间结果中相同 "Key" 的所有 "Value" 进行规约，以得到最终结果。MapReduce 的功能划分使之非常适合用于在大量计算机组成的分布式并行环境中进行数据处理。

④ Tez（支持 DAG 作业的开源计算框架）。Tez 是支持 DAG 作业的开源计算框架，它可以将多个有依赖的作业转换为一个作业从而大幅提升 DAG 作业的性能。Tez 源于 MapReduce 框架，核心思想是将 Map 和 Reduce 两个操作进一步拆分，即 Map 被拆分成 Input、Processor、Sort、Merge 和 Output，Reduce 被拆分成 Input、Shuffle、Sort、Merge、Processor 和 Output 等，这些拆分后的元操作可以灵活组合，产生新的操作，这些新的操作经过一些控制程序组装后，可形成一个大的 DAG 作业。

⑤ Spark（大数据处理通用引擎）。Spark 提供了分布式的内存抽象，其最大的特点就是处理速度快，约是 MapReduce 处理速度的 100 倍。此外，Spark 提供了简单易用的 API，几行代码就能实现字数统计。

⑥ Hive（基于 Hadoop 的数据库）。Hive 定义了一种类似 SQL 的查询语言，将 SQL 语句转换为 MapReduce 任务并在 Hadoop 上执行。Hive 通常用于离线分析。

⑦ Pig（基于 Hadoop 的数据流系统）。Pig 的设计动机是提供一种基于 MapReduce 的即席数据分析工具。人们定义了一种数据流语言 Pig Latin，它可以将脚本转换为 MapReduce 任务并在 Hadoop 上执行。Pig 通常用于进行离线分析。

⑧ Shark（数据分析系统）。Shark 是在集群上进行查询处理和复杂分析的数据分析系统。Shark 扩展了 Hive，提供了面向列存储的机制。Shark 在 Spark 的框架基础上提供和 Hive 一样的 Hive SQL 命令接口，使用 Hive 的 API 来实现查询语法分析、逻辑计划生成。在物理计划执行阶段，Shark 用 Spark 代替 MapReduce，为 SQL 查询和复杂分析函数提供了统一的运行引擎。现在 Shark 已被 Spark 的 Spark SQL 模块所取代。

⑨ Oozie（作业流调度系统）。Oozie 是一个基于工作流引擎的服务器，可以运行 MapReduce 和 Pig 任务，它其实是一个运行在 Java Servlet 容器（如 Tomcat）中的 Java Web 应用。

⑩ HBase（分布式列存储数据库）。HBase 是一个针对结构化数据的可伸缩、高可靠、高性能、分布式和面向列的动态模式数据库。和传统关系数据库不同，HBase 采用了 BigTable 的数据模型——增强的稀疏排序映射表。HBase 能够对大规模数据进行随机、实时读写访问，同时，HBase 中保存的数据可以使用 MapReduce 来进行处理，它将数据存储和并行计算完美地结合在一起。

⑪ ZooKeeper（分布式协调服务）。ZooKeeper 用于解决分布式环境下的数据管理问题，主要包括统一命名、同步状态、管理集群、同步配置等。

⑫ Flume（日志收集系统）。Flume 是开源的日志收集系统，具有分布式、高可靠、高容错、易于定制和扩展的特点。它将数据从产生、传输、处理并最终写入目标路径的过程抽象为数据流，在具体的数据流中，数据源支持在 Flume 中定制数据发送方，从而支持收集不同

协议数据。同时，Flume 数据流具有对日志数据进行简单处理的能力，如过滤、格式转换等。此外，Flume 具有将日志写入多种数据目标的可定制能力。总的来说，Flume 是一个可扩展、适用于复杂环境的海量日志收集系统。

⑬ Sqoop（数据库 TEL 工具）。Sqoop 是 SQL-to-Hadoop 的缩写，主要用于在传统数据库和 Hadoop 之间传输数据。数据的导入和导出本质上利用了 MapReduce 的并行化和容错性。

⑭ Ambari（Hadoop 集群管理、监控的开源工具）。Ambari 是一种基于 Web 的工具，它支持 Hadoop 集群的供应、管理和监控。Ambari 已支持大多数 Hadoop 组件，包括 HDFS、MapReduce、Hive、Pig、HBase、ZooKeeper 和 Sqoop 等。

Hadoop 是一个能够让用户轻松搭建和使用的分布式计算平台。用户可以轻松地在 Hadoop 上开发和运行处理海量数据的应用程序。它主要有以下优点。

- 可靠性高。Hadoop 凭借其按位存储和处理数据的能力，确立了坚实的信任基础，确保了数据管理的可靠性。
- 可扩展性高。Hadoop 允许在可用的计算机集群间分配数据并完成计算任务，集群规模可扩展至数以千计的节点，为数据处理需求的增长提供了扩展空间。
- 处理速度快。Hadoop 能够在节点之间动态地移动数据，并保证各个节点的动态平衡，Hadoop 实现了数据处理的高度并行化，极大地提升了处理速度和效率。
- 容错性高。Hadoop 能够自动保存数据的多个副本，并且能够自动对失败的任务进行重新分配。
- 平台与语言选择灵活。Hadoop 核心框架采用 Java 语言编写，与 Linux 平台的整合紧密，优化了运行环境。同时，它还支持使用其他编程语言（如 C++）开发应用程序，为开发者提供了选择空间，促进了技术栈的多样性和项目的可接入性。

（2）Apache Spark

Apache Spark 是专为大规模数据处理而设计的快速通用计算引擎。Spark 是由美国加州大学伯克利分校的 AMP 实验室开发的开源的类 Hadoop MapReduce 的通用并行框架。Apache Spark 框架如图 1-4 所示。

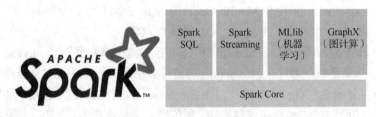

图 1-4　Apache Spark 框架

Spark 框架采用 Scala 语言开发，能够和 Scala 紧密集成，提供类似操作本地数据集的简洁开发体验，让用户能够更高效地进行分布式数据处理，相比 Hadoop 显著提升开发效率与灵活性。

尽管创建 Spark 是为了支持分布式数据集上的迭代作业，但是实际上它是对 Hadoop 的补充，借助第三方集群框架 Mesos 的支持，它可以在 Hadoop 文件系统中并行运行。Spark 具有如下特点。

① 高性能。在进行内存计算时，Spark 的运算速度约是 MapReduce 的 100 倍。

② 易用。Spark 提供了 80 多个高级运算符。

③ 通用。Spark 提供了大量的库，如 Spark SQL、MLlib、GraphX、Spark Streaming 等，开发者可以在同一个应用程序中组合使用这些库。

Spark 的组成如下。

① Spark Core。包含 Spark 的基本功能，定义了弹性分布式数据集（Resilient Distributed Dataset，RDD）及其操作和动作的基本 API，操作是对 RDD 进行的转换，而动作则是对 RDD 执行的操作所触发的计算。其他更高层次的库都是构建在 RDD 和 Spark Core 之上的。

② Spark SQL。提供通过 Apache Hive 的 SQL 变体——Hive 查询语言（HiveQL）与 Spark 进行交互的 API。每个数据库表都被当作一个 RDD，每个 Spark SQL 查询都被转换为 Spark 操作。

③ Spark Streaming。对实时数据流进行处理和控制。Spark Streaming 允许程序像普通 RDD 一样处理实时数据。

④ MLlib。一个常用的机器学习算法库，库中包含可扩展的学习算法，如分类、回归等，这些算法通过 RDD 的 Spark 操作实现。

⑤ GraphX。控制图、并行图操作和计算的一组算法和工具的集合。GraphX 扩展了 RDD API，支持控制图、并行图计算、创建子图、访问路径上的所有顶点等操作。

（3）Apache Storm

Apache Storm（又称 Storm）由 Twitter 开源而来，是自由的开源软件，是分布式的、容错的实时计算系统。Storm 可以非常可靠地处理大规模数据流，常被用于处理 Hadoop 的批量数据。Storm 支持多种编程语言，简单易用。其框架如图 1-5 所示。

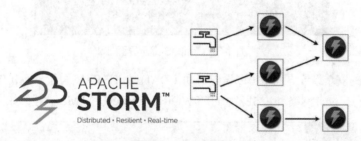

图 1-5　Storm 框架

Storm 应用于多个领域，如实时分析、在线机器学习、不停顿计算、分布式远程过程调用（Remote Procedure Call，RPC）、数据抽取、数据转换和数据加载等。Storm 的处理速度惊人，有数据表明，其每个节点每秒可以处理约 100 万个数据元组。

（4）Apache Flink

Apache Flink 是一个开源、分布式、高性能、高可用的大数据处理引擎，支持实时流（Stream）处理和批（Batch）处理，可部署在多种集群环境中，如 K8s、YARN、Mesos，能够对各种规模的数据进行快速计算。世界各地有很多要求严苛的流处理应用都运行在 Flink 上。

Flink 架构可以分为 4 层，分别为 Deploy 层、Core 层、API 层和 Library 层。

① Deploy 层。关注 Flink 的部署模式，Flink 支持多种部署模式——单机、集群（Standalone/HA/YARN）和云端（GCE/EC2）。

② Core 层。构成 Flink 计算核心，提供支持 Flink 计算的全部核心实现，为 API 层提供基础服务。

③ API 层。提供支持面向无界 Stream 的流处理和面向 Batch 的批处理 API，其中流处理对应 DataStream API，批处理对应 DataSet API。

④ Library 层。也被称为 Flink 应用框架层，是根据 API 层的划分、在 API 层之上构建的、满足特定应用的计算框架，其支持的操作也分为面向流处理和面向批处理两类。面向流处理支持复杂事件处理（Complex Event Processing，CEP）、基于 SQL-like 的操作（基于表关系操作）；面向批处理支持 FlinkML（机器学习库）、Gelly（图计算）、表操作。

任务实施

（1）安装 CentOS（确保 CentOS 版本在 7 及以上，以便配合后续 Docker 的安装）

① 在 VMware 中设置 CentOS 7 镜像，进入 CentOS 提示界面后选择第一项"Install CentOS 7"，如图 1-6 所示。

微课 1-1 Hadoop 环境搭建 1

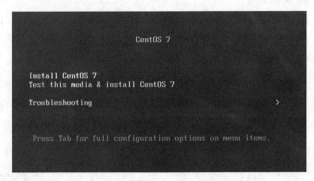

图 1-6 CentOS 提示界面

② 在进入的安装设置界面中可以设置时间（DATE & TIME），分配磁盘（INSTALLATION DESTINATION）、设置网络和主机名（NETWORK & HOST NAME）等，如图 1-7 所示。

图 1-7 安装设置界面

③ 在图 1-7 所示的安装设置界面中单击 "INSTALLATION DESTINATION" 链接，在进入的界面中选择 "I will configure partitioning" 选项，单击 "Done" 按钮，进入分配磁盘空间界面，进行磁盘分配，如图 1-8 所示。

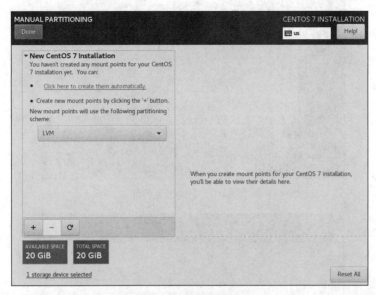

图 1-8　分配磁盘空间界面

在图 1-7 所示的安装设置界面中单击 "NETWORK & HOST NAME" 链接，进入网络和主机名设置界面，进行网络和主机名设置，如图 1-9 所示。

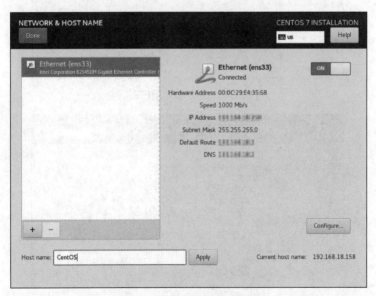

图 1-9　网络和主机名设置界面

④ 单击图 1-7 所示安装设置界面中的 "Begin Installation" 按钮，在进入的账户设置界面中设置 root 用户密码（ROOT PASSWORD）并创建普通用户（USER CREATION），如图 1-10、图 1-11 和图 1-12 所示。

17

图 1-10 账户设置界面

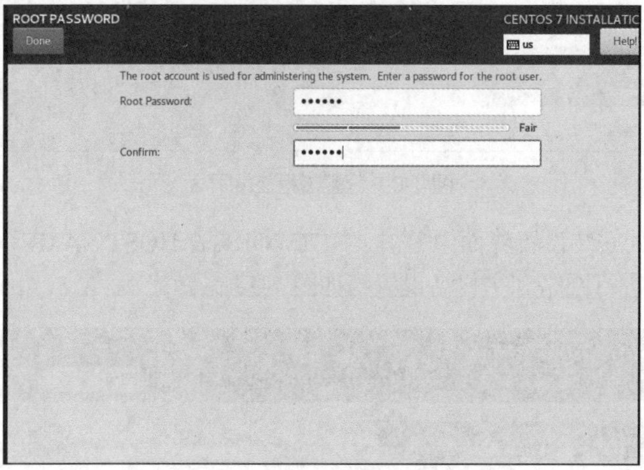

图 1-11 设置 root 用户密码

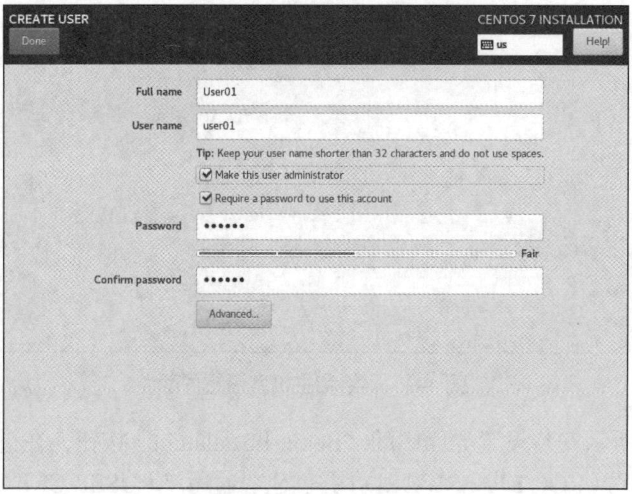

图 1-12 创建普通用户

在创建用户界面中,勾选"Make this user administrator"复选框,可将创建的用户设置为管理员;勾选"Require a password to use this account"复选框,可设置用户登录密码。单击"Done"按钮,完成创建操作。

安装完成后,进入安装完成界面,单击"Reboot"按钮重启操作系统,如图1-13所示。

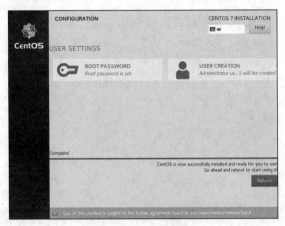

图1-13 安装完成界面

⑤ 进入用户登录界面,在用户登录提示后,输入刚才设置的root用户密码,按"Enter"键,进入CentOS,如图1-14所示。

图1-14 用户登录界面

(2) CentOS中常用命令的使用

① 常用文件和目录操作命令有"cd""ls""ll""mkdir""touch""rm""cp""pwd""mv""vi"等。

"cd"命令用于切换当前工作目录,操作命令及结果如下。

```
[root@CentOS ~]# cd /usr/
[root@CentOS usr]#
```

"ls"命令用于显示指定工作目录中的内容(列出当前工作目录所包含的文件及子目录),操作命令及结果如下。

```
[root@CentOS usr]# ls
bin  etc  games  include  lib  lib64  libexec  local  sbin  share  src  tmp
```

"ll"命令用于显示指定工作目录中的内容(列出当前工作目录所包含的文件及子目录的详细信息,且含隐藏目录),操作命令及结果如下。

```
[root@CentOS usr]# ll
total 116
dr-xr-xr-x.  2 root root 24576 Feb  6 12:33 bin
drwxr-xr-x.  2 root root     6 Apr 11  2018 etc
drwxr-xr-x.  2 root root     6 Apr 11  2018 games
drwxr-xr-x. 35 root root  4096 Feb  5 15:08 include
```

```
dr-xr-xr-x. 29 root root  4096 Feb 25 14:54 lib
dr-xr-xr-x. 41 root root 20480 Feb  5 15:56 lib64
drwxr-xr-x. 23 root root  4096 Feb  5 15:56 libexec
drwxr-xr-x. 15 root root   175 Feb 25 14:55 local
dr-xr-xr-x.  2 root root 16384 Feb  6 12:33 sbin
drwxr-xr-x. 79 root root  4096 Feb  5 15:56 share
drwxr-xr-x.  4 root root    34 Feb  5 12:37 src
lrwxrwxrwx.  1 root root    10 Feb  5 12:37 tmp -> ../var/tmp
```

"mkdir"命令用于建立目录的子目录,操作命令及结果如下。

```
[root@CentOS usr]# mkdir files
[root@CentOS usr]# ls
bin etc files games include lib lib64 libexec local sbin share src tmp
```

"touch"命令用于建立一个新的文件,操作命令及结果如下。

```
[root@CentOS usr]# touch files.txt
[root@CentOS usr]# ls
bin etc files files.txt games include lib lib64 libexec local sbin share src tmp
```

"rm"命令用于删除一个文件,操作命令及结果如下。

```
[root@CentOS usr]# rm files.txt
rm: remove regular empty file 'files.txt'? yes
[root@CentOS usr]# ls
bin etc files games include lib lib64 libexec local sbin share src tmp
```

"rm -r"命令用于删除一个文件或目录(可以包含多个子文件),操作命令及结果如下。

```
[root@CentOS usr]# rm -r files/
rm: remove directory 'files/'? yes
[root@CentOS usr]# ls
bin etc games include lib lib64 libexec local sbin share src tmp
```

"cp"命令用于复制文件或目录,操作命令及结果如下。

```
[root@CentOS usr]# touch files.txt
[root@CentOS usr]# cp files.txt FILE.txt
[root@CentOS usr]# ls
bin etc files.txt FILE.txt games include lib lib64 libexec local sbin share src tmp
```

"pwd"命令用于显示当前工作路径,操作命令及结果如下。

```
[root@CentOS usr]# pwd
/usr
```

"mv"命令用于为文件或目录重命名,或将文件或目录移至其他位置,操作命令及结果如下。

```
[root@CentOS usr]# mv files.txt files02.txt
[root@CentOS usr]# ls
bin etc files02.txt FILE.txt games include lib lib64 libexec local sbin share src tmp
```

"vi"命令用于使用 vi 编辑器打开指定文件。vi 编辑器中的常用参数有"i""I""A""?"":q"":q!"":wq"":wq!"。

参数"i"用于从光标当前所在位置进入编辑状态。

参数"I"用于从光标当前所在行的最前位置进入编辑状态。

参数"A"用于从光标当前所在行的最后位置进入编辑状态。

参数"?"用于查找指定内容所在位置,找到后按"N"键表示向下查找,按"n"键表示向上查找。

参数":q"用于在没有任何修改操作的情况下退出 vi 编辑器。

参数":q!"用于强行退出 vi 编辑器。

参数":wq"用于在进行编辑后保存并退出 vi 编辑器。

参数":wq!"用于在进行编辑后强行保存并退出 vi 编辑器。

② 常用文件权限修改命令有"sudo""chmod""chown"等。

"sudo"命令用于以系统管理员的身份执行指令,操作命令及结果如下。

```
[user01@CentOS usr]$ mkdir student
mkdir: cannot create directory 'student': Permission denied
[user01@CentOS usr]$ sudo mkdir student
[sudo] password for user01:
[user01@CentOS usr]$ ls
bin etc files02.txt FILE.txt games include lib lib64 libexec local sbin
share src student tmp
```

"chmod"命令用于更改指定文件或目录拥有的权限,如将 student 目录的权限更改为 777,操作命令及结果如下。

```
[root@CentOS usr]# chmod 777 student/
[root@CentOS usr]# ll
total 116
dr-xr-xr-x.  2 root root 24576 Feb  6 12:33 bin
drwxr-xr-x. 35 root root  4096 Feb  5 15:08 include
dr-xr-xr-x. 29 root root  4096 Feb 25 14:54 lib
drwxrwxrwx.  2 root root     6 Mar 13 14:40 student
```

"chown"命令用于更改指定文件或目录的拥有者和拥有组,如将 student 目录的拥有者和拥有组从 root 更改为指定的拥有者和拥有组,操作命令及结果如下。

```
[root@CentOS usr]# chown -R user01:user01 student/
[root@CentOS usr]# ll
total 116
dr-xr-xr-x.  2 root   root   24576 Feb  6 12:33 bin
drwxr-xr-x. 35 root   root    4096 Feb  5 15:08 include
dr-xr-xr-x. 29 root   root    4096 Feb 25 14:54 lib
drwxrwxrwx.  2 user01 user01     6 Mar 13 14:40 student
```

③ 常用下载与压缩命令有"tar""yum install"等。

"tar"命令用于解压 tar.gz 类型的压缩文件,操作命令及结果如下。

```
[root@CentOS opt]# tar -zxvf jdk-8u341-linux-x64.tar.gz -C /usr/lib
[root@CentOS lib]# ls
debug     grub           modprobe.d      python2.7        systemd
dracut    jdk1.8.0_341   modules         rpm              tmpfiles.d
```

"yum install"命令用于从指定的服务器自动下载相应软件包，执行完命令后会看到"Complete"字样的提示信息，此处以安装 vim 为例，操作命令如下。

```
[root@CentOS ~]# yum install -y vim
```

可以为系统更换国内 yum 源，先通过"yum install"命令安装软件 wget，操作命令如下。

```
[root@CentOS ~]# yum install -y wget
```

再进入/etc/yum.repo.d 目录，并通过"mv"命令对本地默认 yum 源"CentOS-Base.repo"进行备份，操作命令如下。

```
[root@CentOS yum.repos.d]# mv CentOS-Base.repo CentOS-Base.repo.bak
```

通过 wget 软件执行"wget -O /etc/yum.repos.d/CentOS-Base.repo"命令更换 yum 源，此处以更换阿里云 yum 源为例，操作命令及结果如下。

```
[root@CentOS yum.repos.d]# wget -O /etc/yum.repos.d/CentOS-Base.repo
https://mirrors.aliyun.com/repo/Centos-7.repo
--2023-03-13 06:56:43--  https://mirrors.aliyun.com/repo/Centos-7.repo
Resolving mirrors.aliyun.com (mirrors.aliyun.com)... 218.94.206.184,
222.186.149.91, 222.186.149.101, ...
Connecting to mirrors.aliyun.com (mirrors.aliyun.com)|218.94.206.184|:443...
connected.
HTTP request sent, awaiting response... 200 OK
Length: 2523 (2.5K) [application/octet-stream]
Saving to: '/etc/yum.repos.d/CentOS-Base.repo'
100%[=========================>] 2,523        --.-K/s   in 0.01s
2023-03-13 06:56:43 (220 KB/s) - '/etc/yum.repos.d/CentOS-Base.repo' saved
[2523/2523]
[root@CentOS yum.repos.d]#
```

换源后，执行"yum clean all"命令清除 yum 源缓存，并执行"yum makecache"命令生成缓存，等待提示"Metadata Cache Created"，操作命令及部分结果如下。

```
[root@CentOS yum.repos.d]# yum clean all
Loaded plugins: fastestmirror, ovl
Cleaning repos: base extras updates
Cleaning up list of fastest mirrors
[root@CentOS yum.repos.d]# yum makecache
Loaded plugins: fastestmirror, ovl
Determining fastest mirrors
```

④ 静态 IP 地址设置和远程连接。

通过"ip addr"命令查看自动分配的 IP 地址，此处 IP 地址为 192.168.18.158，如图 1-15 所示。

```
[root@CentOS ~]# ip addr
1: lo: <LOOPBACK,UP,LOWER_UP> mtu 65536 qdisc noqueue state UNKNOWN group default qlen 1000
    link/loopback 00:00:00:00:00:00 brd 00:00:00:00:00:00
    inet 127.0.0.1/8 scope host lo
       valid_lft forever preferred_lft forever
    inet6 ::1/128 scope host
       valid_lft forever preferred_lft forever
2: ens33: <BROADCAST,MULTICAST,UP,LOWER_UP> mtu 1500 qdisc pfifo_fast state UP group default qlen 1000
    link/ether 00:0c:29:e4:35:68 brd ff:ff:ff:ff:ff:ff
    inet 192.168.18.158/24 brd 192.168.18.255 scope global noprefixroute ens33
       valid_lft forever preferred_lft forever
    inet6 fe80::5a35:dbfc:12c5:bb11/64 scope link noprefixroute
       valid_lft forever preferred_lft forever
```

图 1-15 查看自动分配的 IP 地址

修改静态 IP 地址需要执行"vi /etc/sysconfig/network-scripts/ifcfg-ens33"命令，查看 ifcfg-ens33（注意，在操作时，此处的 ifcfg-ens××数字部分可能会略有不同）文件，修改 BOOTPROTO 的值为 static，ONBOOT 的值为 yes，并分别添加 IP 地址、子网掩码、网关、DNS1 和 DNS2。修改完成后，按"Esc"键返回命令行模式，输入":wq!"保存修改并退出，修改后的 ifcfg-ens33 文件内容如下。

```
TYPE="Ethernet"
PROXY_METHOD="none"
BROWSER_ONLY="no"
BOOTPROTO="static"
DEFROUTE="yes"
IPV4_FAILURE_FATAL="no"
IPV6INIT="yes"
IPV6_AUTOCONF="yes"
IPV6_DEFROUTE="yes"
IPV6_FAILURE_FATAL="no"
IPV6_ADDR_GEN_MODE="stable-privacy"
NAME="ens33"
UUID="d1c7d88a-2fdb-4aac-8a08-8ea0cf201d9c"
DEVICE="ens33"
ONBOOT="yes"
IPADDR=192.168.18.158
NETMASK=255.255.255.0
GATEWAY=192.168.18.2
DNS1=192.168.18.2
DNS2=8.8.8.8
```

保存并退出后，重启网络服务，操作命令及结果如下。

```
[root@CentOS lib]# service network restart
Restarting network (via systemctl):                        [  OK  ]
```

在 Windows 上可以通过网络下载 Xshell 等远程登录工具，以便可以远程登录 CentOS。启动远程软件并新建会话，设置会话的名称和要连接的主机 IP 地址，默认连接端口号为 22，如图 1-16 所示。

单击"确定"按钮后，在新进入的界面中进行用户设置，填写要登录的用户名和密码，再次单击"确定"按钮，即可实现远程登录系统，如图 1-17 所示。

图 1-16　会话名称和主机 IP 地址设置

图 1-17　用户设置

任务 1.2　Docker 安装及常用命令

任务描述

（1）学习 Docker 的安装操作。
（2）学习 Docker 的常用命令。

任务目标

（1）了解 Docker 的特点，熟悉 Docker 的组成。
（2）学会 Docker 的安装。
（3）学会 Docker 常用命令的使用。

知识准备

在计算机中，虚拟化是一种资源管理技术，其将计算机的各种实体资源，如服务器、网络适配器、内存及磁盘空间等，进行抽象、转换后再呈现出来，使用户以比原本的组态更好的方式来应用这些资源。这些资源的虚拟部分不受现有资源的架设方式、地域或物理组态所限制。Docker 使用 Go 语言进行开发实现，能够基于 Linux 内核的 Cgroup、Namespace，以及高级联合文件系统（Advanced Union File System，AUFS）类的 UnionFS 等技术，对进程进行封装隔离，属于操作系统层面的虚拟化技术。隔离的进程独立于宿主和其他隔离的进程，因此也称其为容器。Docker 最初是基于 LXC（Linux 原生支持的容器技术，可以提供轻量级的虚拟化）实现的，后来进一步演进为使用 RunC（根据开放容器计划标准创建并运行容器的命令行界面）和 Containerd（工业级标准的容器运行时，它强调简单性、健壮性和可移植性。Containerd 可以在宿主机中管理完整的容器生命周期：容器镜像的传输和存储、容器的执行和管理、存储和网络等）实现。Docker 在容器的基础上，对文件系统、网络互联和进程隔离等，进行了进一步的封装，极大简化了容器的创建和维护。Docker 技术比传统虚拟机技术更为轻便、快捷。

Docker 使用的是"客户端/服务器"架构。Docker 客户端向 Docker 守护进程发送请求，Docker 守护进程处理完所有请求并返回结果，Docker 守护进程负责构建、运行和分发 Docker 容器。Docker 客户端和 Docker 守护进程可以运行在同一个系统上，也可以将 Docker 客户端连接到远程的 Docker 守护进程。Docker 客户端和 Docker 守护进程使用 RESTful API，并通过 UNIX 套接字或网络接口进行通信。Docker 架构如图 1-18 所示。

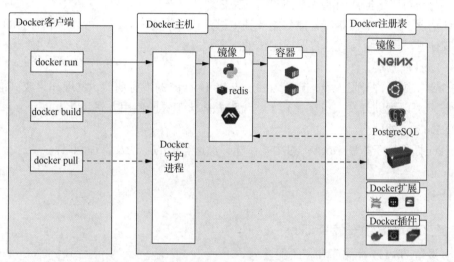

图 1-18 Docker 架构

（1）Docker 守护进程

Docker 守护进程监听 Docker API 请求并管理 Docker 对象，如镜像、容器、网络和卷。Docker 守护进程还可以与其他守护进程通信，以便对 Docker 服务进行管理。

（2）Docker 客户端

Docker 客户端是 Docker 用户与 Docker 互动的主要方式。当使用如"docker run"之类的命令时，Docker 客户端会将这些命令发送给 Docker 守护进程执行。Docker 命令使用 Docker API 实现。Docker 客户端可以与一个以上的 Docker 守护进程通信。

（3）Docker 注册表

Docker 注册表（Docker Hub）存储了 Docker 镜像。Docker Hub 是一个任何人都可以使用的公共注册表，Docker 被配置为默认在 Docker Hub 上寻找镜像。

当使用"docker pull"或"docker run"命令时，所需的镜像会从配置的注册表中拉出，当使用"docker push"命令时，镜像会被推送到配置的注册表中。

（4）镜像

镜像是只读的模板，带有创建 Docker 容器的说明。通常，一个镜像是基于另一个镜像的，并有一些额外的定制。例如，可以建立一个基于 Ubuntu 镜像的镜像，但是要安装 Apache 网络服务器和一些应用程序，以及运行应用程序所需的配置。

用户可以创建自己的镜像，也可以只使用那些由他人创建并发布在注册表上的镜像。如果要创建自己的镜像，则需要创建一个Docker文件，即Dockerfile，并用简单的语法来定义创建镜像和运行它所需的步骤。Dockerfile中的每条指令都会在镜像中创建一个层。当需要修改Dockerfile并重建镜像时，只有那些已经修改的层会被重建。因此，与其他虚拟化技术相比，镜像的使用更加轻巧和快速。

（5）容器

容器是镜像的可运行实例。用户可以使用Docker API或命令行接口（Command Line Interface，CLI）来创建、启动、停止、移动或删除容器，也可以将一个容器连接到一个或多个网络，甚至可以根据其当前状态创建新的镜像。

默认情况下，一个容器与其他容器及其主机是相对隔离的。用户可以控制容器的网络、存储或其他基础子系统与其他容器或主机的隔离程度。

（6）Docker扩展

Docker扩展允许开发人员构建新功能，扩展Docker的现有功能，发现和集成其他工具。Docker扩展可以帮助开发人员优化其工作流程并简化工具的使用方式。

（7）Docker插件

Docker插件用于扩展Docker的功能，能够加速容器的启动及管理网络。

任务实施

1. 在CentOS中安装Docker

（1）设置Docker存储库和相关配置

① 执行"yum"命令安装相关软件包，分别是yum-utils、gcc、gcc-c++，操作命令如下。

```
[root@CentOS ~]# yum install -y yum-utils
[root@CentOS ~]# yum install -y gcc
[root@CentOS ~]# yum install -y gcc-c++
```

② 执行"yum-config-manager --add-repo https://mirrors.aliyun.com/docker-ce/linux/centos/docker-ce.repo"命令，将Docker的镜像仓库设置成国内的镜像仓库，便于后续从仓库中拉取镜像，操作命令及结果如下。

```
[root@CentOS ~]# yum-config-manager --add-repo https://mirrors.aliyun.com/docker-ce/linux/centos/docker-ce.repo

Loaded plugins: fastestmirror
adding repo from: https://mirrors.aliyun.com/docker-ce/linux/centos/docker-ce.repo
grabbing file https://mirrors.aliyun.com/docker-ce/linux/centos/docker-ce.repo to /etc/yum.repos.d/docker-ce.repo
repo saved to /etc/yum.repos.d/docker-ce.repo
```

③ 执行"yum makecache fast"命令更新yum索引，操作命令及结果如下。

```
[root@CentOS ~]# yum makecache fast
Loaded plugins: fastestmirror
Loading mirror speeds from cached hostfile
 * base: mirrors.aliyun.com
 * extras: mirrors.aliyun.com
 * updates: mirrors.aliyun.com
base                                                   | 3.6 kB  00:00:00
docker-ce-stable                                       | 3.5 kB  00:00:00
extras                                                 | 2.9 kB  00:00:00
updates                                                | 2.9 kB  00:00:00
```

④ 执行"yum install docker-ce docker-ce-cli containerd.io docker-compose-plugin"命令安装 docker-ce，等待片刻，Docker 即可安装成功，操作命令及结果如下。

```
[root@CentOS ~]# yum install docker-ce docker-ce-cli containerd.io
docker-compose-plugin
Loaded plugins: fastestmirror
Loading mirror speeds from cached hostfile
```

⑤ 安装成功后，因为 Docker 的运行需要用到各种端口，为避免防火墙的存在对操作过程产生影响，所以在启动 Docker 前，需要先执行"systemctl stop firewalld"命令关闭防火墙，然后执行"systemctl status firewalld"命令查看防火墙的运行状态，如果状态信息显示"inactive (dead)"，则表示防火墙已经关闭，操作命令及结果如下。

```
[root@CentOS ~]# systemctl stop firewalld
[root@CentOS ~]# systemctl status firewalld
● firewalld.service - firewalld - dynamic firewall daemon
   Loaded: loaded (/usr/lib/systemd/system/firewalld.service; enabled; vendor preset: enabled)
   Active: inactive (dead) since Mon 2023-03-13 23:37:59 CST; 17s ago
     Docs: man:firewalld(1)
  Process: 714 ExecStart=/usr/sbin/firewalld --nofork --nopid $FIREWALLD_ARGS (code=exited, status=0/SUCCESS)
 Main PID: 714 (code=exited, status=0/SUCCESS)
```

⑥ 执行"systemctl disable firewalld"命令，禁止防火墙开机自启动，操作命令及结果如下。

```
[root@CentOS ~]# systemctl disable firewalld
Removed symlink /etc/systemd/system/multi-user.target.wants/firewalld.service.
Removed symlink /etc/systemd/system/dbus-org.fedoraproject.FirewallD1.service.
```

⑦ 执行"systemctl start docker"命令来启动 Docker，启动成功后执行"systemctl status docker"命令查看 Docker 的运行状态，如果状态信息显示"active (running)"，则表示 Docker 正常启动，操作命令及结果如下。

```
[root@CentOS ~]# systemctl start docker
[root@CentOS ~]# systemctl status docker
● docker.service - Docker Application Container Engine
   Loaded: loaded (/usr/lib/systemd/system/docker.service; enabled; vendor
```

```
 preset: disabled)
   Active: active (running) since Mon 2023-03-13 15:33:49 CST; 1h 54min ago
     Docs: https://docs.docker.com
 Main PID: 1219 (dockerd)
```

（2）测试 Docker

① 执行 "docker version" 命令查看 Docker 版本，操作命令及结果如下。

```
[root@CentOS ~]# docker version
Client: Docker Engine - Community
 Version:           23.0.0
 API version:       1.42
 Go version:        go1.19.5
 Git commit:        e92dd87
 Built:             Wed Feb  1 17:49:02 2023
 OS/Arch:           linux/amd64
 Context:           default

Server: Docker Engine - Community
 Engine:
  Version:          23.0.0
  API version:      1.42 (minimum version 1.12)
  Go version:       go1.19.5
  Git commit:       d7573ab
  Built:            Wed Feb  1 17:46:49 2023
  OS/Arch:          linux/amd64
  Experimental:     false
 containerd:
  Version:          1.6.16
  GitCommit:        31aa4358a36870b21a992d3ad2bef29e1d693bec
 runc:
  Version:          1.1.4
  GitCommit:        v1.1.4-0-g5fd4c4d
 docker-init:
  Version:          0.19.0
  GitCommit:        de40ad0
```

② 执行 "docker run hello-world" 命令，运行 Docker 的样例测试，如果在运行结果中可以看到 "Hello from Docker!"，则表示 Docker 安装成功，操作命令及结果如下。

```
[root@CentOS ~]# docker run hello-world
Unable to find image 'hello-world:latest' locally
latest: Pulling from library/hello-world
2db29710123e: Pull complete
Digest: 
sha256:2498fce14358aa50ead0cc6c19990fc6ff866ce72aeb5546e1d59caac3d0d60f
```

```
Status: Downloaded newer image for hello-world:latest

Hello from Docker!
```
（3）卸载 Docker

① 执行 "systemctl stop docker" 命令，停止正在运行的 Docker 服务，操作命令及结果如下。

```
[root@CentOS ~]# systemctl stop docker
Warning: Stopping docker.service, but it can still be activated by:
 docker.socket
```

② 执行 "yum remove containerd.io docker-*" 命令，卸载 Docker 及相关软件，操作命令及结果如图 1-19 所示。

图 1-19　卸载 Docker 及相关软件

③ 执行 "rm -rf /var/lib/docker" 命令，删除 Docker 残余文件，执行 "rm -rf /var/lib/containerd" 命令，删除本地已存在的容器文件夹。

学习 Docker 常用命令前，可以将当前普通用户加入 Docker 组里，否则普通用户在使用 Docker 命令时，若不添加或者不使用前置 sudo 命令，则会报错并提示权限不足。添加完成后，当普通用户使用 Docker 命令时，就不用每次都在命令前加上 sudo 了，操作命令及结果如下。

```
[user01@CentOS root]$ docker images
permission denied while trying to connect to the Docker daemon socket at
unix:///var/run/docker.sock: Get "http://%2Fvar%2Frun%2Fdocker.sock/v1.24/
images/json": dial unix /var/run/docker.sock: connect: permission denied
[user01@CentOS root]$ sudo groupadd docker
We trust you have received the usual lecture from the local System
Administrator. It usually boils down to these three things:
   #1) Respect the privacy of others.
   #2) Think before you type.
   #3) With great power comes great responsibility.
[sudo] password for user01:
Sorry, try again.
[sudo] password for user01:
groupadd: group 'docker' already exists
[user01@CentOS root]$ sudo usermod -aG docker $USER
[user01@CentOS root]$ newgrp docker
[user01@CentOS root]$ docker run hello-world
```

```
Hello from Docker!
This message shows that your installation appears to be working correctly.
```

2. Docker 常用命令

（1）Docker 基础命令

启动 Docker 的命令：systemctl start docker。

停止 Docker 的命令：systemctl stop docker。

重启 Docker 的命令：systemctl restart docker。

查看 Docker 状态的命令：systemctl status docker。

设置 Docker 开机自启动的命令：systemctl enable docker。

查看 Docker 概要信息的命令：docker info。

查看 Docker 总体帮助文档的命令：docker --help。

查看 Docker 命令帮助文档的命令：docker 具体命令--help。

（2）Docker 相关镜像操作的命令

① 列出本地主机上镜像的命令。docker images [options]。其中，[options]为可选参数，常用参数有"-a"，表示列出所有本地镜像；"-q"，表示只显示镜像 ID。

操作命令及结果如下。

```
[user01@CentOS root]$ docker images
REPOSITORY     TAG      IMAGE ID        CREATED          SIZE
hello-world    latest   feb5d9fea6a5    17 months ago    13.3kB
[user01@CentOS root]$ docker images -a
REPOSITORY     TAG      IMAGE ID        CREATED          SIZE
hello-world    latest   feb5d9fea6a5    17 months ago    13.3kB
[user01@CentOS root]$ docker images -q
feb5d9fea6a5
[user01@CentOS root]$ docker images -qa
feb5d9fea6a5
```

对"docker images"命令执行结果的解释如表 1-1 所示。

表 1-1 对"docker images"命令执行结果的解释

项目	REPOSITORY	TAG	IMAGE ID	CREATED	SIZE
含义	镜像仓库	镜像的版本标签	镜像 ID	镜像创建时间	镜像大小

TAG 用来代表同一个仓库源中镜像的不同版本，可以使用 REPOSITORY：TAG 来指定镜像的版本，若不指定，则默认拉取最新（latest）版本的镜像。

② 查找 Docker 镜像的命令：docker search [options] 镜像名称。其中，[options]为可选参数，常用参数有"limit N"，表示列出 N 个镜像，默认为 25。使用"docker search"命令查找 Docker 镜像的执行结果如图 1-20 所示。

```
[user01@CentOS root]$ docker search --limit 5 centos
NAME                              DESCRIPTION                                      STARS    OFFICIAL    AUTOMATED
centos                            DEPRECATED; The official build of CentOS.        7538     [OK]
kasmweb/centos-7-desktop          CentOS 7 desktop for Kasm Workspaces             34
bitnami/centos-base-buildpack     Centos base compilation image                    0                    [OK]
bitnami/centos-extras-base                                                         0
couchbase/centos7-systemd         centos7-systemd images with additional debug…    7                    [OK]
```

图 1-20 查找 Docker 镜像的执行结果

对"docker search"命令执行结果的解释如表 1-2 所示。

表 1-2 对"docker search"命令执行结果的解释

项目	NAME	DESCRIPTION	STARS	OFFICIAL	AUTOMATED
含义	镜像名称	镜像说明	点赞数量	是否官方	是否自动创建

③ 下载 Docker 镜像的命令：docker pull 镜像名称:TAG。操作命令及结果如下。

```
[user01@CentOS root]$ docker pull centos:7
7: Pulling from library/centos
2d473b07cdd5: Pull complete
Digest:
sha256:be65f488b7764ad3638f236b7b515b3678369a5124c47b8d32916d6487418ea4
Status: Downloaded newer image for centos:7
docker.io/library/centos:7
```

如果下载镜像时不指定其版本，则会默认下载仓库中最新版本的镜像，即选择标签为 latest。

④ 删除 Docker 镜像的命令如下。

删除一个镜像：docker rmi -f 镜像名称/镜像 ID。

删除多个镜像：docker rmi -f 镜像名称 1:TAG1 镜像名称 2:TAG2 或 docker rmi -f 镜像 ID1 镜像 ID2。多个镜像间使用空格分隔。

删除全部镜像：docker rmi -f $(docker images -qa) 。

删除多个镜像的执行结果如图 1-21 所示。

```
[user01@CentOS root]$ docker rmi -f centos:latest ubuntu:latest
Untagged: centos:latest
Untagged: centos@sha256:a27fd8080b517143cbbbab9dfb7c8571c40d67d534bbdee55bd6c473f432b177
Deleted: sha256:5d0da3dc976460b72c77d94c8a1ad043720b0416bfc16c52c45d4847e53fadb6
Deleted: sha256:74ddd0ec08fa43d09f32636ba91a0a3053b02cb4627c35051aff89f853606b59
Untagged: ubuntu:latest
Untagged: ubuntu@sha256:626ffe58f6e7566e00254b638eb7e0f3b11d4da9675088f4781a50ae288f3322
Deleted: sha256:ba6acccedd2923aee4c2acc6a23780b14ed4b8a5fa4e14e252a23b846df9b6c1
Deleted: sha256:9f54eef412758095c8079ac465d494a2872e02e90bf1fb5f12a1641c0d1bb78b
```

图 1-21 删除多个镜像的执行结果

（3）新建 Docker 容器的命令

新建一个 Docker 容器但不启动的命令：docker create [options] 镜像名称 [COMMAND] [ARG...]。

新建一个 Docker 容器并运行的命令：docker run [options] 镜像名称 [COMMAND] [ARG...]。

其中，[options]的常用参数如下。

-d：后台运行 Docker 容器，并返回容器 ID。

-i：以交互模式运行 Docker 容器，通常与-t 同时使用。

-P：随机端口映射，将 Docker 容器内部端口随机映射到主机端口。

-p：指定端口映射，格式为"主机端口:容器端口"。

-t：为 Docker 容器重新分配一个伪输入终端，通常与-i 同时使用。

--name 名称：指定 Docker 容器的名称。

-h 主机名称：指定 Docker 容器的主机名称。

--network=网桥名：使 Docker 容器加入指定网桥。

操作命令及结果如下。

```
[user01@CentOS root]$ docker run -it -h Centos --name test centos:7 /bin/bash
[root@Centos /]# exit
exit
[user01@CentOS root]$ docker create -it -h Centos01 --name test02 centos:7 /bin/bash
00a37b5567c4051c58239d9595a2884bc31ace1e3a435bda3f8336ff63aac2d0
```

（4）启动和停止 Docker 容器的命令

启动 Docker 容器的命令：docker start 容器 ID/容器名称。

停止 Docker 容器的命令：docker stop 容器 ID/容器名称。

操作命令及结果如下。

```
[user01@CentOS root]$ docker start test
test
[user01@CentOS root]$ docker stop test
test
```

（5）进入和退出 Docker 容器的命令

进入 Docker 容器的第 1 种命令：docker exec -i -t 容器 ID/容器名称/bin/bash。

进入 Docker 容器的第 2 种命令：docker attach 容器 ID/容器名称。

退出 Docker 容器的方式：输入"exit"命令或者按"Ctrl+P+Q"组合键。

以第 1 种命令进入 Docker 容器并使用"exit"命令退出时，不停止 Docker 容器；以方式②进入 Docker 容器并使用"exit"命令退出时，停止 Docker 容器。

操作命令及结果如下。

```
[user01@CentOS root]$ docker exec -i -t test /bin/bash
[root@Centos /]# exit
exit
```

（6）查看和删除 Docker 容器的命令

查看运行中的 Docker 容器的命令：docker ps。

查看所有 Docker 容器的命令：docker ps -a。

操作命令及结果如下。

```
[user01@CentOS root]$ docker ps
CONTAINER ID    IMAGE    COMMAND    CREATED    STATUS    PORTS    NAMES
```

```
f7faef83894e    centos:7        "/bin/bash"     3 minutes ago   Up 58 seconds           test
[user01@CentOS root]$ docker ps -a
CONTAINER ID    IMAGE           COMMAND         CREATED         STATUS          PORTS   NAMES
00a37b5567c4    centos:7        "/bin/bash"     2 minutes ago   Created                 test02
f7faef83894e    centos:7        "/bin/bash"     3 minutes ago   Up About a minute       test
```

删除未启动的 Docker 容器的命令：docker rm 容器 ID/容器名称。

强制删除 Docker 容器的命令：docker rm -f 容器 ID/容器名称。该命令也可用于删除启动中的容器。

操作命令及结果如下。

```
[user01@CentOS root]$ docker rm test02
test02
[user01@CentOS root]$ docker rm -f test
test
```

任务 1.3 Hadoop 环境搭建

任务描述

（1）借助学习论坛、网络视频等网络资源和各种图书资源，学习大数据等相关知识，熟悉 Hadoop 在 3 种运行模式下的安装与配置的异同。

（2）搭建 Hadoop 环境。

任务目标

（1）学会 Hadoop 单机模式的安装与配置。

（2）学会 Hadoop 伪分布式模式的安装与配置。

（3）学会 Hadoop 集群模式的安装与配置。

知识准备

Hadoop 可在 3 种运行模式下搭建，分别为单机模式搭建、伪分布式模式搭建和集群模式搭建。

单机模式即 Hadoop 运行在单一计算机环境中，它不使用分布式文件系统，所有的读写操作都是对本地操作系统的文件系统的直接调用。Hadoop 在单机模式下不会启动 NameNode、DataNode、JobTracker、TaskTracker 等节点，所有事务都运行在一个 Jave 虚拟机（Jave Virtual Machine，JVM）中。在单机模式下，Hadoop 的 Map 和 Reduce 任务作为同一个进程的不同部分来执行。该模式主要用于对 MapReduce 程序的逻辑进行调试，以确保程序正确。

伪分布式模式是在单机上模拟 Hadoop 分布式运行效果，单机上的分布式并不是真正的分布式，而是使用 Java 进程模拟分布式运行中的各类节点，包括 NameNode、DataNode、SecondaryNameNode、JobTracker、TaskTracker。其中，前 3 个节点是从分布式存储的角度来描述的，集群节点由一个 NameNode 和若干个 DataNode 组成，另有一个 SecondaryNameNode 作为 NameNode 的备份；后 2 个节点是从分布式应用的角度来描述的，集群节点由一个 JobTracker 和若干个 TaskTracker 组成，JobTracker 负责任务的调度，TaskTracker 负责并行任务的执行。TaskTracker 必须运行在 DataNode 上，这样便于数据的本地化计算，而 JobTracker 和 NameNode 则无须运行在同一台机器上。Hadoop 本身是无法区分伪分布式和分布式的，因为这两种运行模式下 Hadoop 的配置相似，但其也有区别，伪分布式是在单机上配置的，DataNode 和 NameNode 均位于同一台计算机。

集群模式指 Hadoop 守护进程运行在一个集群上，即使用分布式 Hadoop 时，只有在启动一些守护进程后，才能使用"start-dfs.sh""start-yarn.sh"命令来启动 HDFS 和 YARN。而其他 2 种模式不需要启动这些守护进程。

3 种模式下组件配置的区别如表 1-3 所示。

表 1-3 3 种模式下组件配置的区别

组件名称	属性名称	单机模式	伪分布式模式	集群模式
Common	fs.defaultFS	file:///（默认）	Localhost:9000	Master:9000
HDFS	dfs.replication	N/A	1	3（默认）
MapReduce	mapreduce.framework.name	Local（默认）	YARN	YARN
YARN	yarn.resourcemanager.hostname	N/A	Localhost	Localhost
	yarn.nodemanager.aux_service	N/A	mapreduce_shuffle	mapreduce_shuffle

任务实施

1. 单机模式的安装与配置

（1）安装 JDK

① 在 Xshell 上使用 root 用户登录，在 CentOS 上执行"systemctl start docker"命令启动 Docker，然后执行"systemctl status docker"命令，检查 Docker 的状态，如图 1-22 所示。

```
[root@CentOS ~]# systemctl status docker
● docker.service - Docker Application Container Engine
   Loaded: loaded (/usr/lib/systemd/system/docker.service; enabled; vendor preset: disabled)
   Active: active (running) since Sat 2023-03-25 11:09:01 CST; 1h 19min ago
     Docs: https://docs.docker.com
 Main PID: 1188 (dockerd)
```

图 1-22 检查 Docker 的状态

② 确认 Docker 启动后，执行"docker pull centos:7"命令拉取 CentOS 7 镜像，操作命令及结果如下。

```
[root@CentOS ~]# docker pull centos:7
7: Pulling from library/centos
Digest: sha256:9d4bcbbb213dfd745b58be38b13b996ebb5ac315fe75711bd618426a630e0987
Status: Image is up to date for centos:7
docker.io/library/centos:7
```

项目 1　认识大数据

③ CentOS 7 镜像下载完成后，执行"docker run -dit -h 主机名称 --name 容器名称 --privileged=true centos:7 init"命令，启动并进入 Docker 容器，操作命令及结果如下。

```
[root@CentOS ~]# docker run -dit -h hadoop --name hadoop --privileged=true centos:7 init
b8228a5e9603df7b16714854be992b0f88b2693a9d666fef26bfaabc08377586
```

④ Hadoop 是使用 Java 编写的，所以需要安装 Java 环境。将 CentOS 中准备好的 TAR 包通过"docker cp 宿主机 TAR 包存储路径 容器名称:容器存储路径"命令，将 Jave 开发工具（Java Development Kit，JDK）复制到相应的容器中，操作命令及结果如下。

```
[root@CentOS ~]# docker cp /opt/jdk-8u341-linux-x64.tar.gz hadoop:/opt/
Successfully copied 148.2MB to hadoop:/opt/
[root@CentOS ~]# docker cp /opt/hadoop-2.7.7.tar.gz hadoop:/opt/
Successfully copied 218.7MB to hadoop:/opt/
```

⑤ 进入目标 Docker 容器的存储目录中查看软件包，并执行" tar -zxvf jdk-8u341-linux-x64.tar.gz -C /usr/lib"命令，解压 Java 的 TAR 包，操作命令如下。

```
[root@CentOS ~]# docker exec -it hadoop bash
[root@hadoop /]# cd /opt
[root@hadoop opt]# ls
hadoop-2.7.7.tar.gz  jdk-8u341-linux-x64.tar.gz
[root@hadoop opt]# tar -zxvf jdk-8u341-linux-x64.tar.gz -C /usr/lib
```

解压之后，切换到解压目录，操作命令如下。

```
[root@hadoop /]# cd /usr/lib
```

查看解压文件是否存在，操作命令及结果如下。

```
[root@hadoop lib]# ls
binfmt.d  firewalld  grub     kernel       modules-load.d  python2.7  sse2      tuned
cpp       firmware   jdk1.8.0_341  locale  NetworkManager  rpm        sysctl.d  udev
debug     games      kbd      modprobe.d   os-release      sendmail   systemd   yum-plugins
dracut    gcc        kdump    modules      polkit-1        sendmail.postfix     tmpfiles.d
```

对文件夹进行重命名，操作命令及结果如下。

```
[root@hadoop lib]# mv jdk1.8.0_341 jdk1.8.0
[root@hadoop lib]# ls
binfmt.d  firewalld  grub     kernel       modules-load.d  python2.7  sse2      tuned
cpp       firmware   jdk1.8.0  locale     NetworkManager   rpm        sysctl.d  udev
debug     games      kbd      modprobe.d   os-release      sendmail   systemd   yum-plugins
dracut    gcc        kdump    modules      polkit-1        sendmail.postfix     tmpfiles.d
```

执行"vi ~/.bashrc"命令，编辑用户环境变量，在.bashrc 文件的末尾添加如下配置。

```
export JAVA_HOME=/usr/lib/jdk1.8.0
PATH=$PATH:$JAVA_HOME/bin
```

⑥ 执行"source ~/.bashrc"命令，使新配置的用户环境变量立即生效，并执行"java -version"命令，验证 Java 安装是否成功，操作命令及结果如下。

```
[root@hadoop lib]# source ~/.bashrc
[root@hadoop lib]# java -version
```

```
java version "1.8.0_341"
Java(TM) SE Runtime Environment (build 1.8.0_341-b10)
Java HotSpot(TM) 64-Bit Server VM (build 25.341-b10, mixed mode)
```

（2）安装 Hadoop 2.7.7

① 在 Docker 容器的存储目录中执行 "tar -zxvf hadoop-2.7.7.tar.gz -C /usr/local" 命令，对 Hadoop 进行解压，并切换到解压目录，重命名 hadoop-2.7.7 为 hadoop，操作命令及结果如下。

```
[root@hadoop opt]# tar -zxvf hadoop-2.7.7.tar.gz -C /usr/local
[root@hadoop lib]# cd /usr/local/
[root@hadoop local]# ls
bin  etc  games  hadoop-2.7.7  hbase  include  lib  lib64  libexec  sbin  share  src
[root@hadoop local]# mv hadoop-2.7.7 hadoop
[root@hadoop local]# ls
bin  etc  games  hadoop  hbase  include  lib  lib64  libexec  sbin  share  src
```

② 执行 "vi ~/.bashrc" 命令，配置用户环境变量，在文件的末尾添加如下配置。

```
export HADOOP_HOME=/usr/local/hadoop
PATH=$PATH:$HADOOP_HOME/bin:$HADOOP_HOME/sbin
```

③ 执行 "source ~/.bashrc" 命令，使新配置的用户环境变量立即生效。

（3）配置 Hadoop 环境变量

执行 "cd /usr/local/hadoop/etc/hadoop/" 命令，切换到 hadoop 目录，执行 "vi hadoop-env.sh" 命令，操作命令如下。

```
[root@hadoop local]# cd /usr/local/hadoop/etc/hadoop/
[root@hadoop hadoop]# vi hadoop-env.sh
```

此时，vi 编辑器处于命令行模式，在英文输入状态下，按 "i" 键，切换 vi 编辑器为插入模式，找到 "# export JAVA_HOME=${JAVA_HOME}"，将 "#" 删除，并设置好 JDK 路径，如图 1-23 所示。

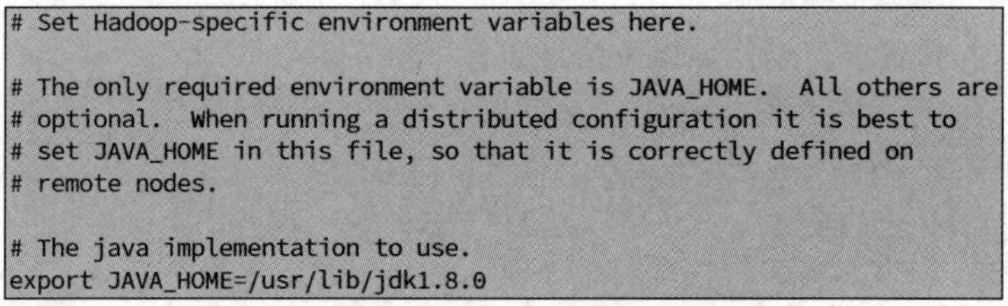

```
# Set Hadoop-specific environment variables here.

# The only required environment variable is JAVA_HOME.  All others are
# optional.  When running a distributed configuration it is best to
# set JAVA_HOME in this file, so that it is correctly defined on
# remote nodes.

# The java implementation to use.
export JAVA_HOME=/usr/lib/jdk1.8.0
```

图 1-23 配置 Hadoop 环境变量

（4）配置 SSH 免密登录

SSH 为 Secure Shell（安全外壳）的缩写，由因特网工程任务组（Internet Engineering Task Force，IETF）的网络小组制定。SSH 协议为建立在应用层基础上的安全协议，专为远程登录会话和其他网络服务提供安全性协议。利用 SSH 协议可以有效地防止远程管理过程中的信息泄露问题产生。

① 执行 "passwd" 命令，为 Docker 容器设置 root 密码，此处密码设置为 "123456"，操作命令及结果如下。

```
[root@hadoop ~]# passwd
Changing password for user root.
New password:
BAD PASSWORD: The password is shorter than 8 characters
Retype new password:
passwd: all authentication tokens updated successfully.
```

② 通过"yum"命令安装 openssh、openssh-server、openssh-clients、net-tools 软件包，操作命令如下。

```
[root@hadoop /]# yum install -y which openssh openssh-server openssh-clients net-tools
```

③ 执行"systemctl start sshd"命令，开启 SSH，操作命令如下。

```
[root@hadoop hadoop]# systemctl start sshd
```

④ 执行"cd ~/.ssh"命令，切换当前路径为 ssh 目录，执行"ssh-keygen -t rsa"命令，生成密钥后，执行"cat id_rsa.pub >> authorized_keys"命令，将公钥追加到 authorized_keys 中，实现 SSH 免密登录配置，操作命令如下。

```
[root@hadoop .ssh]# ssh-keygen -t rsa
[root@hadoop .ssh]# cat id_rsa.pub >> authorized_keys
```

⑤ 执行"ssh localhost"命令，验证配置是否生效，操作命令及结果如下。

```
[root@hadoop .ssh]# ssh localhost
Last login: Mon Mar 13 08:27:43 2023 from localhost
[root@hadoop ~]#
```

（5）验证 Hadoop 单机模式是否安装与配置成功

① 执行"start-all.sh"命令，启动 Hadoop 服务，并通过"jps"命令查看当前进程，操作命令及结果如下。

```
[root@hadoop ~]# start-all.sh
[root@hadoop ~]# jps
906 ResourceManager
1179 NodeManager
1291 Jps
```

可以发现在 Hadoop 单机模式下，进程中并没有 NameNode、DataNode 等守护进程。

执行 Hadoop 自带的 hadoop-mapreduce-examples-2.7.7.jar 程序，验证单机模式配置是否生效。

② 在 HDFS 中创建 input 目录，即执行"hadoop dfs -mkdir input"命令，操作命令及结果如下。

```
[root@hadoop ~]# hadoop dfs -mkdir input
DEPRECATED: Use of this script to execute hdfs command is deprecated.
Instead use the hdfs command for it.
```

在本地目录中创建一个 wordcount.txt 文件，该文件内容可以自行设定。注意，单词间需要用"空格"隔开，操作命令如下。

```
[root@hadoop ~]# vi wordcount.txt
```

本书中，为 wordcount.txt 文件编辑如下内容。

```
hello hadoop
hello java
hello world
```

将编辑好的 wordcount.txt 文件保存并退出，执行"hadoop dfs -copyFromLocal /usr/local/hadoop/wordcount.txt input"命令，将其上传到 HDFS 的 input 目录中，操作命令及结果如下。

```
[root@hadoop ~]# hadoop dfs -copyFromLocal /usr/local/hadoop/wordcount.txt input
DEPRECATED: Use of this script to execute hdfs command is deprecated.
Instead use the hdfs command for it.
```

③ 这里以运行 hadoop-mapreduce-examples-2.7.7.jar 为例进行说明。执行"hadoop jar /usr/local/hadoop/share/hadoop/mapreduce/hadoop-mapreduce-examples-2.7.7.jar wordcount input output"命令，使用 Hadoop 自带的例子 hadoop-mapreduce-examples-2.7.7.jar，在 input 中以空格分割统计字词的个数，操作命令如下。

```
[root@hadoop ~]# hadoop jar /usr/local/hadoop/share/hadoop/mapreduce/hadoop-mapreduce-examples-2.7.7.jar wordcount input output
```

因为这些操作是在 root 用户的主目录中进行的，所以可以直接看到生成的 output 目录。可以看到，output 目录中生成了 2 个文件，通过"cat ./part-r-00000"命令查看文件内容，操作命令及结果如下。

```
[root@hadoop ~]# ls
anaconda-ks.cfg  input  output  wordcount.txt
[root@hadoop ~]# cd output/
[root@hadoop output]# ls
_SUCCESS  part-r-00000
[root@hadoop output]# cat ./part-r-00000
hadoop  1
hello   3
java    1
world   1
```

程序执行成功，说明 Hadoop 单机模式的安装与配置成功。

2. 伪分布式模式的安装与配置

（1）创建有端口映射的 Docker 容器

在宿主机上执行"docker run -dit -h master --name master -p 50070:50070 --privileged=true centos:7 init"命令，运行 Docker 容器，操作命令如下。

```
[root@CentOS ~]# docker run -dit -h master --name master -p 50070:50070 --privileged=true centos:7 init
```

微课 1-2 Hadoop 环境搭建 2

（2）设置密码，安装 JDK、Hadoop 以及相关软件

设置密码，通过"yum"命令安装 openssh、openssh-clients、openssh-server、net-tools、which 软件包。随后，安装 JDK、Hadoop 并配置环境变量（步骤参考 Hadoop 单机模式的安装与配置）。

（3）配置 SSH 免密登录

① 在当前目录中执行"systemctl start sshd"命令，开启 SSH，执行"ssh-keygen -t rsa"命令，生成密钥对，执行"ssh-copy-id master"命令，将公钥复制到 master 的 authorized_keys 文件中，操作命令及结果如下。

```
[root@master /]# systemctl start sshd
[root@master /]# ssh-keygen -t rsa
[root@master /]# ssh-copy-id master
```

② 切换到~/.ssh 目录，执行"ls"命令，查看生成的认证文件，操作命令及结果如下。

```
[root@master /]# cd ~/.ssh
[root@master .ssh]# ls
authorized_keys  id_rsa  id_rsa.pub  known_hosts
```

③ 验证 master 免密登录，操作命令及结果如下。

```
[root@master .ssh]# ssh master
[root@master ~]# exit
logout
Connection to master closed.
```

（4）配置 Hadoop

① 进入/usr/local/hadoop/etc/hadoop 中，执行"vi core-site.xml"命令，并切换到编辑模式，配置 core-site.xml 文件，此处注意"<value>/usr/local/hadoop/data/tmp</value>"配置项中的目录需要手动创建。配置结束后，按"Esc"键，退出编辑模式，在输入":wq"后按"Enter"键保存并退出。需要编辑的 core-site.xml 文件内容如下。

```xml
<configuration>
<property>
<name>fs.defaultFS</name>
<value>hdfs://localhost:9000</value>
</property>
<property>
<name>hadoop.tmp.dir</name>
<value>/usr/local/hadoop/data/tmp</value>
</property>
</configuration>
```

② 进入/usr/local/hadoop/etc/hadoop 目录，执行"vi hdfs-site.xml"命令，并切换到编辑模式，配置 hdfs-site.xml 文件，此处注意"<value>/usr/local/hadoop/data/tmp/name</value>"和"<value>/usr/local/hadoop/data/tmp/data</value>"配置项中的目录需要手动创建。配置结束后，按"Esc"键，退出编辑模式，输入":wq"后按"Enter"键保存并退出。需要编辑的 hdfs-site.xml 文件内容如下。

```xml
<configuration>
<property>
<name>dfs.replication</name>
<value>1</value>
</property>
```

```xml
<property>
<name>dfs.namenode.name.dir</name>
<value>/usr/local/hadoop/data/tmp/name</value>
</property>
<property>
<name>dfs.datanode.name.dir</name>
<value>/usr/local/hadoop/data/tmp/data</value>
</property>
</configuration>
```

③ 进入/usr/local/hadoop/etc/hadoop 目录，复制 mapred-site.xml.template 文件，创建新文件 mapred-site.xml 后，执行"vi mapred-site.xml"命令，并切换到编辑模式，配置 mapred-site.xml 文件，配置结束后，输入":wq"后按"Enter"键保存并退出，操作命令如下。

```
[root@master hadoop]# cp mapred-site.xml.template mapred-site.xml
[root@master hadoop]# vi mapred-site.xml
```

mapred-site.xml 文件内容如下。

```xml
<configuration>
<property>
<name>mapreduce.framework.name</name>
<value>yarn</value>
</property>
</configuration>
```

④ 进入/usr/local/hadoop/etc/hadoop 目录，执行"vi yarn-site.xml"命令，并切换到编辑模式，配置 yarn-site.xml 文件，配置结束后，输入":wq"后按"Enter"键保存并退出。需要编辑的 yarn-site.xml 文件内容如下。

```xml
<configuration>
<!-- Site specific YARN configuration properties -->
<property>
<name>yarn.nodemanager.aux_services</name>
<value>mapreduce_shuffle</value>
</property>
<property>
<name>yarn.resourcemanager.hostname</name>
<value>localhost</value>
</property>
</configuration>
```

（5）格式化 HDFS

切换到 Hadoop 的安装目录，执行"./bin/hdfs namenode -format"命令，格式化节点，如果在返回的信息中看到"Exiting with status 0"，则表示格式化节点成功，操作命令及结果如下。

```
[root@master hadoop]# ./bin/hdfs namenode -format
23/03/13 12:03:18 INFO util.ExitUtil: Exiting with status 0
23/03/13 12:03:18 INFO namenode.NameNode: SHUTDOWN_MSG:
```

```
/*************************************************
SHUTDOWN_MSG: Shutting down NameNode at master/172.17.0.2
*************************************************/
```

（6）验证测试

① 在任意目录中，执行"start-all.sh"命令，启动 Hadoop 集群，操作命令如下。

```
[root@master hadoop]# start-all.sh
```

② 执行"jps"命令查看进程，操作命令及结果如下。

```
[root@master hadoop]# jps
257 NameNode
921 NodeManager
381 DataNode
1086 Jps
543 SecondaryNameNode
703 ResourceManager
```

与单机模式相比，可以发现在伪分布式模式进程中有 NameNode、DataNode 等守护进程。

③ 测试 HDFS 和 YARN。在 Windows 浏览器地址栏中输入"http://docker 宿主机 IP 地址:50070"，进入 HDFS 信息界面，如图 1-24 所示。

图 1-24　HDFS 信息界面（1）

如果可以成功显示该界面，则说明 Hadoop 伪分布式模式的安装与配置成功。

3. 集群模式的安装与配置

在 Hadoop 集群模式的安装与配置中，Hadoop 集群架构如图 1-25 所示。

如果之前已经运行过伪分布式模式，建议在切换到集群模式之前，先删除之前在伪分布式模式下生成的临时文件。

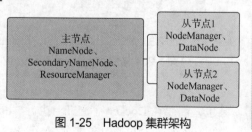

图 1-25　Hadoop 集群架构

微课 1-3　Hadoop 环境搭建 3

（1）创建网桥

为了使后续 Hadoop 集群规划在同一个网段里，需要执行"docker network create --subnet=172.18.0.0/24 --gateway=172.18.0.1 网桥名称"命令来创建自定义网桥，操作命令及结果如下。

```
[root@CentOS ~]# docker network create --subnet=172.18.0.0/24
--gateway=172.18.0.1 hadoop
1d4d5fde4ad0d08ab50afa1fdf4eb10302541d66a04562c7036c48c2cf2b6b56
```

（2）创建并运行容器

① 需要创建 3 个容器，实现一主二从的集群配置，执行"docker run -dit -h master1 --name master1 --network=网桥名称 --ip=172.18.0.2 --privileged=true -p 50070:50070 -p 8088:8088 centos:7 init"命令创建 master1 容器。然后退出容器并执行"docker run -dit -h slave1 --name slave1 --network=网桥名称 --ip=172.18.0.3（或--ip=172.18.0.4）--privileged=true -p 8089:8088（或 8090:8088）centos:7 init"命令创建 slave1 容器，创建 slave2 容器的步骤与之类似，不再赘述，操作命令及结果如下。

创建 master1。

```
[root@CentOS ~]# docker run -dit -h master1 --name master1 --network=hadoop
--ip=172.18.0.2 --privileged=true -p 50070:50070 -p 8088:8088 centos:7 init
```

创建 slave1。

```
[root@CentOS ~]# docker run -dit -h slave1 --name slave1 --network=hadoop
--ip=172.18.0.3 --privileged=true -p 8089:8088 centos:7 init
```

创建 slave2。

```
[root@CentOS ~]# docker run -dit -h slave2 --name slave2 --network=hadoop
--ip=172.18.0.4 --privileged=true -p 8090:8088 centos:7 init
```

② 3 个容器都创建完成后，查看它们是否已加入网桥，执行"docker network inspect 网桥名称"命令查看网桥中 3 个容器的 IP 地址，如图 1-26 所示。

```
"ConfigOnly": false,
"Containers": {
    "1a3ed0b818d63fd1902bae9fd0b292d1e5178332844860e6cde1b18fea314053": {
        "Name": "slave1",
        "EndpointID": "dd1098210365c1d91f88ead46d549ce06391b18300ae4c57e2246dcf77d7a5b4",
        "MacAddress": "02:42:ac:12:00:03",
        "IPv4Address": "172.18.0.3/16",
        "IPv6Address": ""
    },
    "26fbe6d70f785e5c04dda8a7cc44c976999e30546f72f138209601ad1b9bf422": {
        "Name": "master1",
        "EndpointID": "b383443e3257e0da76d34ee19fd91cb322dee9ec9b5e2128aa5fc2c2005df46a",
        "MacAddress": "02:42:ac:12:00:02",
        "IPv4Address": "172.18.0.2/16",
        "IPv6Address": ""
    },
    "c3811fde5c00ce3dc1851506674020af5284a0910b47ed974a1d32b29c9a918f": {
        "Name": "slave2",
        "EndpointID": "1395e941c5b14db31dd87a9d18f47cff2c98c13d747715758b71373b2615050c",
        "MacAddress": "02:42:ac:12:00:04",
        "IPv4Address": "172.18.0.4/16",
        "IPv6Address": ""
    }
},
"Options": {},
"Labels": {}
```

图 1-26　查看网桥中 3 个容器的 IP 地址

（3）基础配置

① 传输所需安装包，需要在 Docker 宿主机上执行"docker cp /opt/×××.tar.gz 容器名称:目标存储路径"命令（×××为安装包名称），操作命令及结果如下。

```
[root@CentOS ~]# docker cp /opt/jdk-8u341-linux-x64.tar.gz master1:/opt/
Successfully copied 148.2MB to master1:/opt/
[root@CentOS ~]# docker cp /opt/hadoop-2.7.7.tar.gz master1:/opt/
Successfully copied 218.7MB to master1:/opt/
```

② 执行"docker exec -it master1 bash"命令，为master1开启一个新的Bash Shell会话，在master1中先执行"passwd"命令设置密码，再通过"yum"命令安装所需的软件包，需要安装的件包有net-tools、openssh、openssh-server、openssh-clients、vim、which。安装完成后，执行"systemctl start sshd"命令开启SSH服务，操作命令及结果如下。

```
[root@CentOS ~]# docker exec -it master1 bash
[root@master1 /]# passwd
Changing password for user root.
New password:
BAD PASSWORD: The password is shorter than 8 characters
Retype new password:
passwd: all authentication tokens updated successfully.
[root@master1 /]# yum install -y net-tools openssh openssh-server openssh-clients vim which
[root@master1 /]# systemctl start sshd
```

③ 在slave1和slave2中重复上述操作中的第②步，完成设置。

（4）测试互通性

测试集群中各个节点之间的互通性（这里使用"ping"命令进行测试），操作命令及结果如下。

```
[root@master1 /]# ping slave2
PING slave2 (172.18.0.4) 56(84) bytes of data.
64 bytes from slave2.hadoop (172.18.0.4): icmp_seq=1 ttl=64 time=0.054 ms
64 bytes from slave2.hadoop (172.18.0.4): icmp_seq=2 ttl=64 time=0.050 ms
64 bytes from slave2.hadoop (172.18.0.4): icmp_seq=3 ttl=64 time=0.056 ms
```

随后，在集群的各个节点上配置SSH免密登录，实现主（Master）节点免密登录各个从（Slave）节点。

（5）主节点的配置

① 进入~/.ssh目录，执行"ssh-keygen -t rsa"命令，生成密钥对，如图1-27所示。

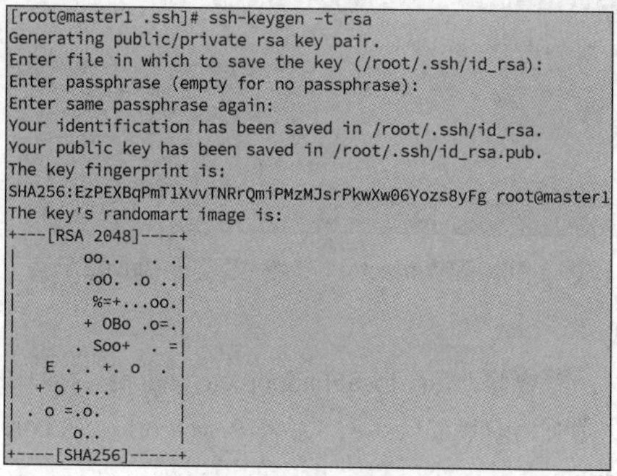

```
[root@master1 .ssh]# ssh-keygen -t rsa
Generating public/private rsa key pair.
Enter file in which to save the key (/root/.ssh/id_rsa):
Enter passphrase (empty for no passphrase):
Enter same passphrase again:
Your identification has been saved in /root/.ssh/id_rsa.
Your public key has been saved in /root/.ssh/id_rsa.pub.
The key fingerprint is:
SHA256:EzPEXBqPmT1XvvTNRrQmiPMzMJsrPkwXw06Yozs8yFg root@master1
The key's randomart image is:
+---[RSA 2048]----+
|        oo.. . .|
|       .oO. .o .|
|      .%=+...oo.|
|       + OBo .o=.|
|      . Soo+  . =|
|       E . .+. o |
|      . + o +... |
|      . o =.o.   |
|         o..     |
+----[SHA256]-----+
```

图1-27　生成密钥对

② 执行"ssh-copy-id master1"命令，验证SSH免密登录master1成功，如图1-28所示。

```
[root@master1 .ssh]# ssh-copy-id master1
/usr/bin/ssh-copy-id: INFO: Source of key(s) to be installed: "/root/.ssh/id_rsa.pub"
The authenticity of host 'master1 (172.18.0.2)' can't be established.
ECDSA key fingerprint is SHA256:p5Vth2W2xlPqEvL1i6ImAggNs9ssebWULkEgWbibSA0.
ECDSA key fingerprint is MD5:10:b8:f7:19:9d:63:90:5e:62:df:68:28:d8:3f:54:94.
Are you sure you want to continue connecting (yes/no)? yes
/usr/bin/ssh-copy-id: INFO: attempting to log in with the new key(s), to filter out any that are already installed
/usr/bin/ssh-copy-id: INFO: 1 key(s) remain to be installed -- if you are prompted now it is to install the new keys
root@master1's password:

Number of key(s) added: 1

Now try logging into the machine, with:   "ssh 'master1'"
and check to make sure that only the key(s) you wanted were added.

[root@master1 .ssh]# ssh master1
Last login: Sat Mar 25 11:09:54 2023 from 192.168.18.1
[root@master1 ~]# exit
logout
Connection to master1 closed.
```

图 1-28　验证 SSH 免密登录 master1 成功

（6）从节点的配置

在各从节点上重复 master1 配置操作，其中，需要注意的是，执行 "ssh-copy-id hostname" 命令时，需要更换 hostname 为 "slave1" 或 "slave2"，即 hostname 为当前主机的主机名。

进入 master1 节点的 ~/.ssh 目录，执行 "ssh-copy-id slave1" 和 "ssh-copy-id slave2" 命令，完成主节点登录各从节点的免密配置，如图 1-29 和图 1-30 所示。

```
[root@master1 .ssh]# ssh-copy-id slave1
/usr/bin/ssh-copy-id: INFO: Source of key(s) to be installed: "/root/.ssh/id_rsa.pub"
The authenticity of host 'slave1 (172.18.0.3)' can't be established.
ECDSA key fingerprint is SHA256:Y6pT7uzLOJWjTTPl3FRUVNcDer0z2NdQfVWnhmj69zA.
ECDSA key fingerprint is MD5:45:c3:ba:4b:9a:4f:37:69:69:80:2c:c1:bd:a9:8f:c6.
Are you sure you want to continue connecting (yes/no)? yes
/usr/bin/ssh-copy-id: INFO: attempting to log in with the new key(s), to filter out any that are already installed
/usr/bin/ssh-copy-id: INFO: 1 key(s) remain to be installed -- if you are prompted now it is to install the new keys
root@slave1's password:

Number of key(s) added: 1

Now try logging into the machine, with:   "ssh 'slave1'"
and check to make sure that only the key(s) you wanted were added.
```

图 1-29　配置 master1 节点免密登录 slave1 节点

```
[root@master1 .ssh]# ssh-copy-id slave2
/usr/bin/ssh-copy-id: INFO: Source of key(s) to be installed: "/root/.ssh/id_rsa.pub"
The authenticity of host 'slave2 (172.18.0.4)' can't be established.
ECDSA key fingerprint is SHA256:T1UcW1qa49h7PLo9Gz2KwJjGP2ljtML0/znBEVivY44.
ECDSA key fingerprint is MD5:bd:c4:68:69:81:a8:63:5b:87:61:95:a2:56:f9:db:f9.
Are you sure you want to continue connecting (yes/no)? yes
/usr/bin/ssh-copy-id: INFO: attempting to log in with the new key(s), to filter out any that are already installed
/usr/bin/ssh-copy-id: INFO: 1 key(s) remain to be installed -- if you are prompted now it is to install the new keys
root@slave2's password:

Number of key(s) added: 1

Now try logging into the machine, with:   "ssh 'slave2'"
and check to make sure that only the key(s) you wanted were added.
```

图 1-30　配置 master1 节点免密登录 slave2 节点

（7）配置集群模式

在配置集群模式时，需要修改 /usr/local/hadoop/etc/hadoop 目录中的配置文件，这里仅设置正常启动所必需的设置项，包括 slaves（3.×版本名为 workers）、core-site.xml、hdfs-site.xml、mapred-site.xml、yarn-site.xml 共 5 个文件，更多的设置项可查看官方说明文档。

① 修改 slaves 文件，将主节点仅作为 NameNode 使用，将 slaves 文件中原来的 localhost 删除，并添加两个从节点的主机名，操作命令如下。

```
[root@master1 hadoop]# vim slaves
```
添加内容如下。
```
slave1
slave2
```
② 编辑 core-site.xml 文件，此处注意 "<value>/usr/local/hadoop/data/tmp</value>" 配置项中的目录需要手动创建，内容如下。
```
<configuration>
<property>
<name>fs.defaultFS</name>
<value>hdfs://master1:9000</value>
</property>
<property>
<name>hadoop.tmp.dir</name>
<value>/usr/local/hadoop/data/tmp</value>
<description>Abase for other temporary directories.</description>
</property>
</configuration>
```
③ 编辑 hdfs-site.xml 文件，此处注意 "<value>/usr/local/hadoop/data/tmp/name</value>" 和 "<value>/usr/local/hadoop/data/tmp/data</value>" 配置项中的目录需要手动创建，内容如下。
```
<configuration>
<property>
<name>dfs.namenode.secondary.http-address</name>
<value>master1:50090</value>
</property>
<property>
<name>dfs.replication</name>
<value>1</value>
</property>
<property>
<name>dfs.namenode.name.dir</name>
<value>/usr/local/hadoop/data/tmp/name</value>
</property>
<property>
<name>dfs.datanode.data.dir</name>
<value>/usr/local/hadoop/data/tmp/data</value>
</property>
</configuration>
```
④ 编辑 mapred-site.xml 文件，内容如下。
```
<configuration>
<property>
<name>mapreduce.framework.name</name>
```

```xml
<value>yarn</value>
</property>
<property>
<name>mapreduce.jobhistory.address</name>
<value>master1:10020</value>
</property>
<property>
<name>mapreduce.jobhistory.webapp.address</name>
<value>master1:19888</value>
</property>
</configuration>
```

⑤ 编辑 yarn-site.xml 文件，内容如下。

```xml
<configuration>
<!-- Site specific YARN configuration properties -->
<property>
<name>yarn.resourcemanager.hostname</name>
<value>master1</value>
</property>
<property>
<name>yarn.nodemanager.aux-services</name>
<value>mapreduce_shuffle</value>
</property>
</configuration>
```

上述文件全部配置完成后，需要把 master1 节点上的/usr/local/hadoop 目录、/usr/lib/jdk1.8.0 目录以及 ~/.bashrc 文件分别复制到各个从节点上。要注意，复制完 ~/.bashrc 文件后，在各个从节点上务必执行 "source ~/.bashrc" 命令，以便新的环境变量能够立即生效。

在 master1、slave1 和 slave2 节点上需要执行的一系列操作命令如下。

```
[root@master1 local]# scp -r ~/.bashrc slave1: ~/.bashrc
[root@master1 local]# scp -r ~/.bashrc slave2: ~/.bashrc
[root@master1 local]# scp -r /usr/local/hadoop/ slave1:/usr/local
[root@master1 local]# scp -r /usr/local/hadoop/ slave2:/usr/local
[root@master1 local]# scp -r /usr/lib/jdk1.8.0 slave1:/usr/lib
[root@master1 local]# scp -r /usr/lib/jdk1.8.0 slave2:/usr/lib
[root@slave1 ~]# source ~/.bashrc
[root@slave2 ~]# source ~/.bashrc
```

（8）验证测试

① 首次启动 Hadoop 集群时，需要在 master1 节点上执行 NameNode 节点的格式化操作，执行 "hdfs namenode -format" 命令，完成格式化操作，操作命令如下。

```
[root@master1 local]# hdfs namenode -format
```

② 启动 Hadoop 集群，在 master1 节点上执行 "start-all.sh" 命令，随后依次在各个节点上执行 "jps" 命令，查看各个节点的进程运行情况，操作命令及结果如下。

```
[root@master1 local]# start-all.sh
[root@master1 local]# jps
897 SecondaryNameNode
1057 ResourceManager
706 NameNode
1319 Jps
[root@slave1 ~]# jps
373 DataNode
586 Jps
475 NodeManager
[root@slave2 /]# jps
548 Jps
437 NodeManager
335 DataNode
```

在 Windows 浏览器地址栏中输入"http://Docker 宿主机 IP 地址:50070",进入 HDFS 信息界面,如图 1-31 所示。

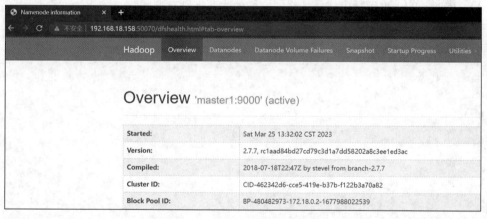

图 1-31　HDFS 信息界面(2)

在 Windows 浏览器地址栏中输入"http://Docker 宿主机 IP 地址:8088",打开 Web 控制台,可以查看集群状态。如果 DataNode 中没有配置 yarn-site.xml 文件,则在网页中无法查看节点信息;如果配置了 yarn-site.xml 文件,则在网页中可以查看节点信息,如图 1-32 所示。

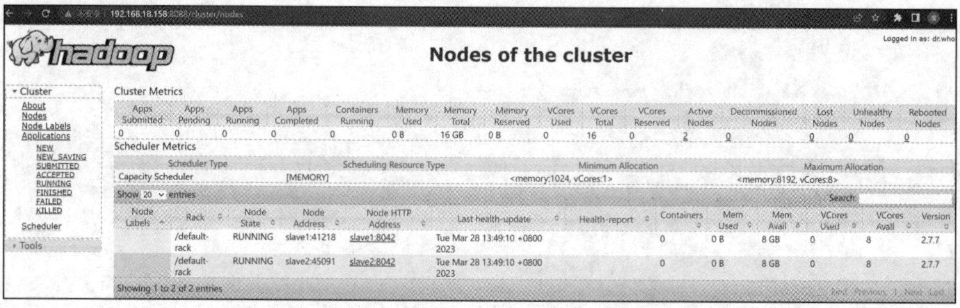

图 1-32　查看节点信息

项目小结

本项目主要介绍了大数据的定义、大数据的基本特征、大数据处理与分析的相关技术、工具或产品等，完成了 CentOS 的安装、Hadoop 的安装与配置，以及 Docker 的安装和常用命令的学习等。学完本项目，读者可对大数据的相关技术、应用、产业链以及大数据的分析流程、工具等建立较为清晰的认知，能够了解使用大数据的意义，为建立大数据思维奠定良好的基础。

课后练习

1. 请简述 Docker 的常用命令及其作用。
2. 本地的镜像文件存放在哪些目录中？如何更改 Docker 的默认存储设置？
3. Docker 的配置文件存放在哪个目录中？
4. Hadoop 的 3 种运行模式是什么？它们的主要区别有哪些？
5. 请简述 Hadoop 的组件。
6. Hadoop 2.×的守护进程有哪些？

项目 2 Hive 环境搭建与基本操作

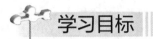

【知识目标】
了解 Hive 产生的背景、Hive 架构。
识记 Hive SQL 常用语句。

【素质目标】
具有严谨细致的工作态度和工作作风。
具有良好的团队协作意识和业务沟通能力。
具有良好的表达能力和文档查阅能力。
具有规范的编程意识和较好的数据洞察力。

【技能目标】
学会 Hive 的安装与配置。
学会 Hive SQL 的基本操作。

项目描述

Hadoop 的 MapReduce 程序擅长处理大规模数据集。其使用对于程序员而言比较容易，但对于非专业者而言则有一定难度。Hive 最初就是为分析和处理海量日志而生的，通过 Hive，用户无须了解 MapReduce 也可以方便地进行大数据处理。Hive 提供了丰富的 SQL 查询方式，使用户能方便地对存储在 Hadoop 分布式文件系统中的数据进行分析，Hive 将结构化的数据文件映射为一张数据库表。除此之外，Hive 提供的完整 SQL 查询功能，可以将 SQL 语句转换为 MapReduce 任务来运行，简化了复杂的数据处理工作。

本项目主要引导学生完成 Hive 的安装与配置，并学习 Hive SQL 的基本操作。

任务 2.1 Hive 的安装与配置

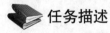

（1）学习 Hive 的相关技术知识，了解 Hive 的产生背景、特点和 Hive SQL 的基本操作等。
（2）完成 Hive 的安装与配置。

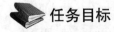

（1）熟悉 Hive 的特点。

（2）学会 MySQL 的安装与配置。
（3）学会 Hive 的安装与配置。

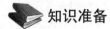 知识准备

1. Hive 数据结构

Hive 中所有的数据都存储在 HDFS 中，Hive 中包含以下数据结构。

（1）表（Table）

Hive 中的表和数据库中的表在概念上是类似的，每一个表在 Hive 中都有一个相应的目录存储数据。

（2）分区（Partition，可选）

在 Hive 中，表中的每个分区都对应一个目录，所有的分区的数据都存储在对应的目录中。

（3）桶（Bucket，可选）

桶对指定列计算其哈希值，并根据列的哈希值将数据散列到不同的桶中，其目的是进行并行处理，使每一个桶对应一个文件。

2. Hive 架构

Hive 架构如图 2-1 所示。

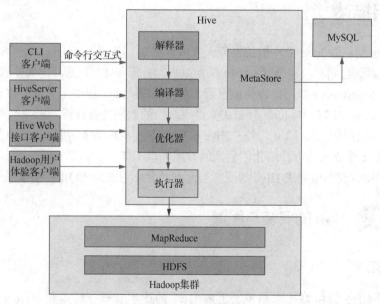

图 2-1　Hive 架构

HDFS 和 MapReduce 是 Hive 架构的基础。用户接口主要有 CLI 客户端、HiveServer 客户端、Hive Web 接口（Hive Web Interface，HWI）客户端和 Hadoop 用户体验（Hadoop User Experience，HUE）客户端，其中常用的是 CLI 客户端。在 CLI 客户端启动时，会同时启动一个 Hive 副本。在 Windows 中，可通过 Java 数据库互连（Java Database Connectivity，JDBC）连接 HiveServer 客户端的图形界面工具，包括 SQuirrel SQL Client、Oracle SQL Developer 及

项目 2 Hive 环境搭建与基本操作

DbVisualizer。HWI 客户端是 Hive CLI 的一个 Web 替换方案，它通过浏览器访问 Hive，通过 Web 控制台与 Hadoop 集群进行交互来分析及处理数据。HUE 客户端是一个开源的 Hadoop UI 系统，通过 HUE 客户端，用户可以在浏览器端的 Web 控制台上与 Hadoop 集群进行交互来分析处理数据，如运行 MapReduce 作业，执行 Hive SQL 语句等。MetaStore 用于存储和管理 Hive 的元数据，它使用关系数据库（MySQL、Derby 等）来保存元数据，Hive 中的元数据包括表的名称、表的列和分区及其属性、表的属性（是否为外部表等）、表的数据所在目录等。Hive 通过解释器、编译器、优化器和执行器完成 Hive SQL 查询语句从词法分析、语法分析、编译、优化到查询计划的生成。生成的查询计划存储在 HDFS 中，随后由 MapReduce 调用。大部分的查询、计算都由 MapReduce 来完成。

3. Hive 与传统关系数据库的对比

使用 Hive 的命令行接口很像操作关系数据库，但是 Hive 和关系数据库有很大的不同，具体如下。

（1）宏观角度

① Hive 和关系数据库存储文件的系统不同。Hive 使用的是 Hadoop 的 HDFS，关系数据库使用的是服务器本地的文件系统。

② 计算模型不同。Hive 使用的计算模型是 MapReduce，而关系数据库使用的是自己设计的计算模型。

③ 应用场景不同。关系数据库是为实时查询业务而设计的，实时性较好；而 Hive 是为对海量数据进行数据挖掘而设计的，实时性较差。

④ 存储能力与计算能力不同。Hive 架构的基础包括 Hadoop，因此很容易扩展自己的存储能力和计算能力，而关系数据库在此方面表现较差。

（2）微观角度

① 数据加载模式不同。在关系数据库中，表的加载模式（数据库存储数据的文件格式）是在数据加载时强制确定的，如果加载数据时发现加载的数据不符合模式，则关系数据库会拒绝加载数据，这种模式称为"写时模式"，写时模式会在数据加载时对数据格式进行检查校验。和关系数据库的加载过程不同，Hive 在加载数据时不会对数据进行检查，也不会更改被加载的数据文件，检查数据格式的操作是在查询时执行的，这种模式称为"读时模式"。在实际应用中，写时模式在加载数据时会对列进行索引，对数据进行压缩，因此加载数据的速度很慢，但是当数据加载好后，查询数据的速度很快。当面临非结构化数据及未知存储模式的场景的挑战时，传统数据库处理起来会相对吃力，而 Hive 则能够展现出其独特的价值与优势。

② 对具体数据的操作能力不同。关系数据库有一个重要的特点——其可以对某一行或某些行的数据进行更新、删除操作，而 Hive 不支持对某具体行的操作，Hive 对数据的操作只支持覆盖原数据和追加数据。更新、事务和索引都是关系数据库的特点，这些 Hive 都不支持，也不打算支持，原因是 Hive 的架构是为对海量数据进行处理而设计的，全数据的扫描是常态，针对某些具体数据进行操作的效率是很差的。对于更新操作，Hive 是通过查询操作，对原表的数据进行转换后存储在新表中的，这和关系数据库的更新操作有很大不同。

4. Hive 的执行流程

Hive 的执行流程如图 2-2 所示。

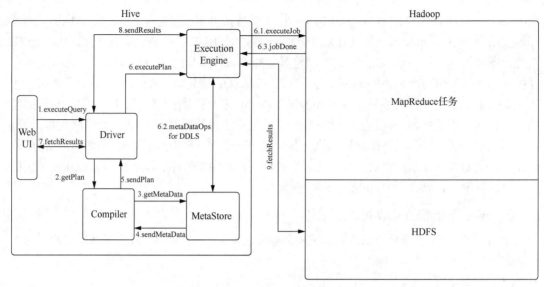

图 2-2　Hive 的执行流程

（1）executeQuery

Hive 界面包含命令行和 Web 用户界面（Web User Interface，Web UI），它将查询（Query）语句发送到驱动（Driver，指数据库驱动程序，如 JDBC、ODBC 等）中执行。

（2）getPlan

Driver 通过编译器（Compiler）解析 Query 语句，验证 Query 语句的语法、查询计划或者查询条件。

（3）getMetaData

Compiler 将元数据（MetaData）请求发送给元数据存储（MetaStore）。

（4）sendMetaData

MetaStore 将 MetaData 作为响应发送给 Compiler。

（5）sendPlan

Compiler 检查要求并重新发送 Driver 的计划。此时，对 Query 语句的解析和编译完成。

（6）executePlan

Driver 将执行计划发送到执行引擎（Execution Engine），该过程的执行流程如下。

① executeJob。Hadoop 内部执行的是 MapReduce 任务，Execution Engine 发送一个任务到 Hadoop，并在 Hadoop 中执行 MapReduce 任务。

② metaDataOps for DDLS。Execution Engine 在执行 executeJob 操作发送任务的同时，对 Hive 的 MetaData 进行相应操作。

③ jobDone。Hadoop 将 MapReduce 任务的操作结果发送到 Execution Engine。

（7）fetchResults

Hive 的 Web UI 从 Driver 中提取操作结果。

（8）sendResults

Execution Engine 发送操作结果到 Driver。

（9）fetchResults

Execution Engine 从 Hadoop 的 HDFS 读取临时文件的内容。

项目 2 Hive 环境搭建与基本操作

📚 **任务实施**

Hive 工具中默认使用的是 Derby 数据库，该数据库使用简单、操作灵活，但是存在一定的局限性。Hive 支持使用第三方数据库（MySQL 等），通过配置可以把 MySQL 集成到 Hive 工具中，MySQL 功能更强大，企业应用也更广泛。

微课 2-1 Hive 的安装与配置

1. MySQL 的安装与配置

（1）准备 MySQL 安装包并安装依赖软件

在 Docker 宿主机上将 MySQL 安装包复制到容器内，操作命令及结果如下。

```
[root@CentOS opt]# docker cp /opt/mysql-5.7.23-linux-glibc2.12-x86_64.tar.gz master1:/opt
[root@CentOS opt]# docker cp mysql-connector-java-5.1.40.tar.gz master1:/opt
[root@CentOS opt]# docker start master1
master1
[root@CentOS opt]# docker exec -it master1 bash
[root@master1 /]# yum install -y initscripts
[root@master1 simple]# yum install -y libaio
[root@master1 simple]# yum -y install numactl
```

（2）安装 MySQL

① 创建 simple 目录，并进入安装包所在目录，通过 "ll" 命令查看上一步复制到容器内的安装包，操作命令及结果如下。

```
[root@master1 /]# mkdir simple
[root@master1 opt]# ll
total 1082072
-rw-r--r--. 1 root root  92834839 Feb 17 03:27 apache-hive-1.2.1-bin.tar.gz
-rw-r--r--. 1 root root 218720521 Feb  5 16:03 hadoop-2.7.7.tar.gz
-rw-r--r--. 1 root root 148162542 Feb  5 16:03 jdk-8u341-linux-x64.tar.gz
-rw-r--r--. 1 root root 644399365 Feb 15 15:55 mysql-5.7.23-linux-glibc2.12-x86_64.tar.gz
-rw-r--r--. 1 root root   3911557 Feb 15 15:55 mysql-connector-java-5.1.40.tar.gz
```

② 解压 MySQL 安装包到 simple 目录，将解压后的目录重命名为 mysql（重命名是为了后续操作方便），操作命令及结果如下。

```
[root@master1 opt]# tar -zxvf mysql-5.7.23-linux-glibc2.12-x86_64.tar.gz -C /simple
[root@master1 opt]# cd /simple
[root@master1 simple]# mv mysql-5.7.23-linux-glibc2.12-x86_64 mysql
```

（3）配置环境变量

① 执行 "vim ~/.bashrc" 命令，在 .bashrc 文件的末尾增加如下配置。

```
export MYSQL_HOME=/simple/mysql
PATH=$PATH:$MYSQL_HOME/bin
```

53

② 执行"source ~/.bashrc"命令,使得环境变量立即生效,操作命令如下。
[root@master1 simple]# source ~/.bashrc

(4) 用户及权限相关配置

① 执行"groupadd mysql""useradd -r -g mysql mysql"命令,添加mysql组和mysql用户,操作命令如下。
[root@master1 simple]# groupadd mysql
[root@master1 simple]# useradd -r -g mysql mysql

② 执行"chgrp -R mysql"和"chown -R mysql"命令,进入mysql目录,并更改其所属的组和用户,操作命令如下。
[root@master1 simple]# cd mysql/
[root@master1 mysql]# chgrp -R mysql
[root@master1 mysql]# chown -R mysql

③ 执行mysql_install_db脚本,对MySQL中的data目录进行初始化。注意,MySQL服务进程mysqld运行时会访问data目录,所以必须由启动mysqld进程的用户(即之前设置的mysql用户)执行此脚本,或者由root用户执行此脚本。执行命令时应加上参数--user=mysql,启动过程中会生成密码,将密码复制并保存好,第一次登录时需要使用该密码,操作命令及结果如下。
[root@master1 simple]# mysqld --initialize --user=mysql basedir=/simple/mysql --datadir=/simple/mysql/data
2023-02-17T02:15:32.370477Z 0 [Warning] TIMESTAMP with implicit DEFAULT value is deprecated. Please use --explicit_defaults_for_timestamp server option (see documentation for more details).
2023-02-17T02:15:32.568321Z 0 [Warning] InnoDB: New log files created, LSN=45790
2023-02-17T02:15:32.596016Z 0 [Warning] InnoDB: Creating foreign key constraint system tables.
2023-02-17T02:15:32.649874Z 0 [Warning] No existing UUID has been found, so we assume that this is the first time that this server has been started. Generating a new UUID: f4dfe2b7-ae68-11ed-aa2e-0242ac120002.
2023-02-17T02:15:32.650615Z 0 [Warning] Gtid table is not ready to be used. Table 'mysql.gtid_executed' cannot be opened.
2023-02-17T02:15:32.651801Z 1 [Note] A temporary password is generated for root@localhost: tZYvmX:zD1)j

最后一条信息中的"tZYvmX:zD1)j"是登录MySQL的初始密码。

④ 在mysql目录中,除了data目录,将其余目录和文件的所有者修改为root用户,mysql用户只需作为mysql/data目录中所有文件的所有者即可,操作命令如下。
[root@master1 simple]# cd mysql/
[root@master1 mysql]# chown -R root .
[root@master1 mysql]# chown -R mysql data

(5) 配置启动文件

① 为了使MySQL服务的重启操作更加便携,避免手动切换到目录bin并执行./mysqld_safe --user=mysql命令,可以对启动文件进行相应配置,以便能够通过执行"/etc/init.d/mysql.server start"命令进行启动,操作命令如下。

项目 2　Hive 环境搭建与基本操作

```
[root@master1 mysql]# cp support-files/mysql.server /etc/init.d/mysql
```

②进入/etc/init.d 目录，编辑 mysql 文件，操作命令如下。

```
[root@master1 mysql]# cd /etc/init.d/
[root@master1 init.d]# vim mysql
```

修改 mysql 文件，在文件中修改 2 个目录的位置，具体修改内容如下。

```
basedir=/simple/mysql
datadir=/simple/mysql/data
```

③启动 MySQL 服务，操作命令及结果如下。

```
[root@master1 /]# service mysql start
Starting MySQL.Logging to '/simple/mysql/data/master1.err'.
 SUCCESS!
```

④执行"mysql -uroot -p"命令，登录 MySQL，提示输入密码，将刚才复制的密码粘贴到冒号后面，按"Enter"键，进入 MySQL 命令行模式，操作命令及结果如下。

```
[root@master1 /]# mysql -uroot -p
Enter password:
Welcome to the MySQL monitor.  Commands end with ; or \g.
Your MySQL connection id is 2
Server version: 5.7.23

Copyright (c) 2000, 2018, Oracle and/or its affiliates. All rights reserved.

Oracle is a registered trademark of Oracle Corporation and/or its
affiliates. Other names may be trademarks of their respective
owners.

Type 'help;' or '\h' for help. Type '\c' to clear the current input statement.

mysql>
```

进入命令行模式后，为了方便登录，可以修改密码，这里执行"set password for 'root'@'localhost'=password('123456');"命令，将密码修改为"123456"（也可以根据需要自行设定），操作命令及结果如下。

```
mysql> set password for 'root'@'localhost'=password('123456');
Query OK, 0 rows affected, 1 warning (0.00 sec)
```

（6）任务测试

①重新启动 MySQL 服务并进行登录（密码已修改为"123456"），操作命令及结果如下。

```
[root@master1 /]# mysql -uroot -p
Enter password:
Welcome to the MySQL monitor.  Commands end with ; or \g.
Your MySQL connection id is 3
Server version: 5.7.23 MySQL Community Server (GPL)

Copyright (c) 2000, 2018, Oracle and/or its affiliates. All rights reserved.

Oracle is a registered trademark of Oracle Corporation and/or its
affiliates. Other names may be trademarks of their respective
owners.

Type 'help;' or '\h' for help. Type '\c' to clear the current input statement.

mysql>
```

如果登录时忘记了密码，则需要重新设定密码。在 MySQL 服务的停止状态下，执行"mysqld_safe --user=mysql --skip-grant-tables --skip-networking &"命令跳过密码验证，打开一个新的窗口，此时可以免密进入 MySQL 的命令行模式，操作命令及结果如下。

```
[root@master1 ~]# mysqld_safe --user=mysql --skip-grant-tables --skip-networking &
[1] 4990
[root@master1 ~]# 2023-03-25T07:43:03.687388Z mysqld_safe Logging to '/simple/mysql/data/master1.err'.
2023-03-25T07:43:03.704239Z mysqld_safe Starting mysqld daemon with databases from /simple/mysql/data
```

免密登录 MySQL，操作命令及结果如下。

```
[root@master1 ~]# mysql -u root
Welcome to the MySQL monitor.  Commands end with ; or \g.
Your MySQL connection id is 2
Server version: 5.7.23-log MySQL Community Server (GPL)
Copyright (c) 2000, 2018, Oracle and/or its affiliates. All rights reserved.
Oracle is a registered trademark of Oracle Corporation and/or its affiliates. Other names may be trademarks of their respective owners.
Type 'help;' or '\h' for help. Type '\c' to clear the current input statement.
mysql>
```

② 进入 MySQL 命令行模式后创建数据库（提供给 Hive 使用），操作命令及结果如下。

```
mysql> create database myhive;
Query OK, 1 row affected (0.01 sec)
```

③ 在命令行模式下，执行"use mysql"命令，使用 mysql 数据库，执行"UPDATE user SET authentication_string=   PASSWORD('123456') where USER='root';"命令，重置密码为 123456，执行"flush privileges;"命令刷新权限，操作命令及结果如下。

```
mysql> use mysql
mysql> UPDATE user SET authentication_string=PASSWORD('123456') where USER='root';
Query OK, 1 row affected, 1 warning (0.01 sec)
Rows matched: 1  Changed: 1  Warnings: 1

mysql> flush privileges;
Query OK, 0 rows affected (0.01 sec)
```

2. 基于 HDFS 和 MySQL 的 Hive 环境搭建

（1）解压 Hive

进入存放 Hive 安装包的目录，执行"tar –zxvf apache-hive-1.2.1-bin.tar.gz –C /simple"命令，把 Hive 安装包解压到/simple 目录中。执行完解压命令后，在/simple 目录中可以查看到解压后的目录"apache-hive-1.2.1-bin"，修改目录名称为"hive1.2.1"，操作命令及结果如下。

```
[root@master1 opt]# tar -zxvf apache-hive-1.2.1-bin.tar.gz -C /simple
[root@master1 opt]# cd /simple
```

项目 2　Hive 环境搭建与基本操作

```
[root@master1 simple]# ll
total 0
drwxr-xr-x.  8 root root  159 Feb 17 06:33 apache-hive-1.2.1-bin
drwxr-xr-x. 10 root mysql 141 Feb 17 02:09 mysql
[root@master1 simple]# mv apache-hive-1.2.1-bin hive1.2.1
```

（2）配置 Hive

① 解压 mysql-connector-java-5.1.40.tar.gz，进入解压目录，将驱动文件 mysql-connector-java-5.1.40-bin.jar 复制到 hive1.2.1 目录的 lib 目录下，操作命令如下。

```
[root@master1 opt]# tar -zxvf mysql-connector-java-5.1.40.tar.gz
[root@master1 opt]# cd mysql-connector-java-5.1.40
[root@master1mysql-connector-java-5.1.40]#cp
mysql-connector-java-5.1.40-bin.jar /simple/hive1.2.1/lib
```

② 切换到/simple/hive1.2.1/conf 目录，执行 "cp hive-env.sh.template ./hive-env.sh" 命令，利用配置文件模板复制并生成配置文件 hive-env.sh，并查看所生成的文件，操作命令及结果如下。

```
[root@master1 opt]# cd /simple/hive1.2.1/conf
[root@master1 conf]# cp hive-env.sh.template ./hive-env.sh
[root@master1 conf]# ll
-rw-rw-r--. 1 root root   1139 Apr 29 2015 beeline-log4j.properties.template
-rw-rw-r--. 1 root root 168431 Jun 19 2015 hive-default.xml.template
-rw-r--r--. 1 root root   2378 Feb 17 06:38 hive-env.sh
-rw-rw-r--. 1 root root   2378 Apr 29 2015 hive-env.sh.template
-rw-rw-r--. 1 root root   2662 Apr 29 2015 hive-exec-log4j.properties.template
-rw-rw-r--. 1 root root   3050 Apr 29 2015 hive-log4j.properties.template
-rw-rw-r--. 1 root root   1593 Apr 29 2015 ivysettings.xml
```

编辑配置文件 hive-env.sh，修改 "HADOOP_HOME" 项为 "/usr/local/hadoop"，具体操作如下。

```
HADOOP_HOME=/usr/local/hadoop
```

③ 切换到/simple/hive1.2.1/conf 目录，执行 "mv hive-default.xml.template hive-site.xml" 命令，将 hive-default.xml.template 文件重命名为 hive-site.xml，操作命令如下。

```
[root@master1 conf]# mv hive-default.xml.template hive-site.xml
```

编辑 hive-site.xml 文件的内容（建议删除原有文件中的所有内容后再进行编辑），具体内容如下。

```
<configuration>
<property>
        <name>javax.jdo.option.ConnectionURL</name>

<value>jdbc:mysql://localhost:3306/myhive?createDatabaseIfNoExist=true&amp;useSSL=false</value>
</property>
<property>
```

```xml
        <name>javax.jdo.option.ConnectionDriverName</name>
        <value>com.mysql.jdbc.Driver</value>
</property>
<property>
        <name>javax.jdo.option.ConnectionUserName</name>
        <value>root</value>
</property>
<property>
        <name>javax.jdo.option.ConnectionPassword</name>
        <value>123456(root密码)</value>
</property>
</configuration>
```

④ 操作结束后，在目录$HIVE_HOME/bin下修改文件hive-config.sh，添加如下内容。

```
# Allow alternate conf dir location.
HIVE_CONF_DIR="${HIVE_CONF_DIR:-$HIVE_HOME/conf}"
export HIVE_CONF_DIR=$HIVE_CONF_DIR
export HIVE_AUX_JARS_PATH=$HIVE_AUX_JARS_PATH
# Default to use 256MB
export HADOOP_HEAPSIZE=${HADOOP_HEAPSIZE:-256}
export JAVA_HOME=/usr/lib/jdk1.8.0
export HIVE_HOME=/simple/hive1.2.1
export HADOOP_HOME=/usr/local/hadoop
```

⑤ 编辑环境变量文件~/.bashrc，在文件的末尾添加如下内容。

```
export MYSQL_HOME=/simple/mysql
export HIVE_HOME=/simple/hive1.2.1
PATH=$PATH:$MYSQL_HOME/bin:$HIVE_HOME/bin
```

执行"source ~/.bashrc"命令，使配置的环境变量立即生效。

⑥ 执行"start-all.sh"命令，启动Hadoop服务后，再执行"./hive"命令进入Hive Shell环境，操作命令及结果如下。

```
[root@master1 bin]# ./hive
Logging initialized using configuration in jar:file:/simple/hive1.2.1/lib/hive-common-1.2.1.jar!/hive-log4j.properties
hive>
```

若正确进入Hive Shell环境，则表示Hive安装与配置成功。

 Hive的应用

任务描述

学习Hive SQL的基本操作、Hive中函数的使用方法，以及分区表和分桶的相关操作方法。

项目 2　Hive 环境搭建与基本操作

任务目标

（1）熟悉 Hive SQL 的基本操作。
（2）学会 Hive 中函数的使用方法以及分区表和分桶的创建方法。

知识准备

1. 创建表

使用 Hive 创建表的语法格式如下。

```
CREATE [TEMPORARY] [EXTERNAL] TABLE [IF NOT EXISTS]
[db_name.]table_name
[(col_name data_type[COMMENT col_comment],...)]
[COMMENT table_comment]
[PARTITIONED BY (col_name data_type[COMMENT col_comment],...)]
[CLUSTERED BY （col_name,col_name,...)[SORTED BY
(col_name[ASC|DESC],...)] INTO num_buckets BUCKETS]
[SKEWED BY (col_name,col_name,...)]
ON ((col_value,col_value,...),(col_value,col_value,...),...)
[STORED AS DIRECTORIES]
[
[ROW FORMAT row_format]
[STORED AS file_format]
| STORED BY 'store.handler.class.name'[WITH SERDEPROPERTIES(...)]
]
[LOCATION hdfs_path]
[TBLPROPERTIES (property_name=property_value,...)]
[AS select_statement];
```

CREATE TABLE 用于创建一张指定名称的表。如果名称相同的表已经存在，则抛出异常；可以用 IF NOT EXISTS 选项来忽略这个异常。

EXTERNAL 关键字可以让用户创建一个外部表，在创建表的同时指定一个指向实际数据的路径（LOCATION）。Hive 创建内部表时，会将数据移动到数据库指向的路径；Hive 创建外部表时，仅记录数据所在的路径，不对数据的位置做任何改变。在删除表的时候，内部表的元数据和数据会被一起删除，而外部表只删除元数据，不删除数据。

Hive 创建表时，会通过自定义的 SerDe 或使用 Hive 内置的 SerDe 类型指定数据的序列化和反序列化方式。在创建表的时候，还需要为表指定列，在指定列的同时会指定自定义的 SerDe，Hive 通过 SerDe 确定表的具体列的数据。如果没有指定 ROW FORMAT 或者 ROW FORMAT DELIMITED，则会使用内置的 SerDe。

存储格式指定为 STORED AS SEQUENCEFILE|TEXTFILE|RCFILE。如果文件数据是纯文本数据，则可以使用 STORED AS TEXTFILE 的存储格式，也可以使用更高级的存储格式，如 ORC、PARQUET 等。

创建内部表的语句如下。
```
create table emp(
empno int,ename string,job string,mgr int,hiredate string,sal
double,comm double,deptno int)row format delimited fields terminated by '\t';
```
创建外部表的语句如下。
```
create external table emp_external(
empno int,ename string,job string,mgr int,hiredate string,sal double,comm
double,deptno int)row format delimited fields terminated by '\t' location
'/hive_external/emp/';
```

2. 修改表

Hive 中的修改表操作包括重命名表、添加列、更新列等。

下面对 Hive 中的修改表操作进行说明。

```
//重命名表操作
ALTER TABLE table_name RENAME TO new_table_name;
//将 emp 表重命名为 emp_new
ALTER TABLE emp RENAME TO emp_new;
//添加、更新列操作
ALTER TABLE table_name ADD|REPLACE COLUMNS (col_name data_type
[COMMENT col_comment],...);
//创建测试表
create table student(id int,age int,name string) row format delimited fields
terminated by '\t';
//添加一列 address
alter table student add columns(address string);
//更新所有的列
alter table student replace columns(id int,name string);
```

3. 查看 Hive 数据库和表的相关信息

下面对查看 Hive 数据库和表的相关信息的操作进行说明。

```
//查看所有数据库
show databases;
//查看数据库中的表
show tables;
//查看表的所有分区信息
show partitions table_name;
//查看 Hive 支持的所有函数
show functions;
//查看表的信息
desc extended t_name;
//查看更加详细的表信息
desc formatted table_name;
```

4. 使用 LOAD 语句将文本文件的数据加载到 Hive 表中

LOAD 语句的语法格式如下。

```
LOAD DATA [LOCAL] INPATH 'filepath' [OVERWRITE] INTO
TABLE tablename [PARTITION (partcol1=val1,partcol2=val2,...)];
```

LOAD 只是单纯的复制、移动操作,表示将数据文件复制或移动到 Hive 表对应的位置。filepath 可以是相对路径,也可以是绝对路径。如果指定了 LOCAL 关键字,则 LOAD 语句会查找本地文件系统中的 filepath;如果没有指定 LOCAL 关键字,则根据 INPATH 中的统一资源标识符(Uniform Resource Identifier,URI)查找文件,此时需包含 filepath 的完整 URI。如果使用了 OVERWRITE 关键字,则目标表(或者分区)中的内容会被删除,并将 filepath 指向的文件/目录中的内容加载到表(或者分区)中。如果目标表(或者分区)中已经有一个文件,并且文件名和 filepath 中的文件名相同,那么原有的文件会被新文件所覆盖。

下面通过实例说明 LOAD 语句的相关操作。

```
//将本地文件加载到 Hive 表中
LOAD DATA LOCAL INPATH '/home/hadoop/data/emp.txt' INTO TABLE emp;
//将 HDFS 文件加载到 Hive 表中
LOAD DATA INPATH '/data/hive/emp.txt' INTO TABLE emp;
//OVERWRITE 关键字的使用,将本地文件加载到 Hive 表中,会覆盖表中已有的数据
LOAD DATA LOCAL INPATH '/home/hadoop/data/emp.txt' OVERWRITE INTO TABLE emp;
//OVERWRITE 关键字的使用,将数据加载到 Hive 分区中,会覆盖表中已有的数据
LOAD DATA LOCAL INPATH '/home/hadoop/data/order.txt' OVERWRITE INTO TABLE
order_partition PARTITION(event_month='2014-05');
```

5. 使用 INSERT 语句将查询结果插入 Hive 表中

INSERT 语句的语法格式如下。

```
INSERT OVERWRITE TABLE tablename1 [PARTITION
(partcol1=val1,partcol2=val2 ...)] select_statement1 FROM from_statement;
```

创建原始数据表,用于测试 INSERT 语句的相关操作,具体如下。

```
DROP TABLE order_4_partition;
CREATE TABLE order_4_partition(
orderNumber STRING,
event_time STRING
)
ROW FORMAT DELIMITED FIELDS TERMINATED BY '\t';
LOAD DATA LOCAL INPATH '/home/hadoop/data/order.txt' OVERWRITE INTO TABLE
ORDER_4_PARTITION;
INSERT OVERWRITE TABLE ORDER_PARTITION PARTITION(event_month='2017-07')SELECT *
FROM order_4_partition;
```

下面通过实例说明 INSERT 语句的相关操作。

```
//使用 INSERT 语句将查询结果插入 Hive 表中
INSERT INTO TABLE account SELECT id,age,name FROM account_tmp;
//复制原表的指定字段
```

```sql
create table emp2 as select empno,ename,job,deptno from emp;
//使用 INSERT 语句将结果插入 Hive 表中
insert into table order_partition partition(event_month='2017-7')
select * from order_4_partition;
```

6. 使用 INSERT 语句将 Hive 表中的数据导出到文件系统

使用 INSERT 语句将 Hive 表中的数据导出到文件系统的语法格式如下。

```
INSERT OVERWRITE [LOCAL] DIRECTORY directory1 SELECT ... FROM ...
```

INSERT 语句的操作实例说明如下。

```sql
//使用 INSERT 语句将数据导出到本地
INSERT OVERWRITE LOCAL directory '/home/hadoop/hivetmp'
ROW FORMAT DELIMITED FIELDS TERMINATED BY '\t' LINES
TERMINATED BY '\n'select * from emp;
//使用 INSERT 语句将数据导出到 HDFS 中
INSERT OVERWRITE directory '/hivetmp/' select * from emp;
```

7. 使用 SELECT 语句进行查询操作

SELECT 语句的语法格式如下。

```
SELECT[ALL | DISTINCT]select_expr,select_expr,...
FROM table_reference
[WHERE where_condition]
[GROUP BY col_list [HAVING condition]]
[CLUSTER BY col_list
| [DISTRIBUTE BY col_list] [SORT BY| ORDER BY col_list]
]
[LIMIT number]
```

（1）ORDER BY 会对输入数据进行全局排序，因此，当只有一个 Reducer 操作时，会由于输入规模较大导致计算时间较长。

（2）SORT BY 会在输入数据进入 Reducer 操作前完成对其的排序操作。因此，如果使用 SORT BY 进行排序，并且设置 mapred.reduce.tasks>1，则 SORT BY 只能保证每个 Reducer 操作的输出是有序的，而不能保证全局有序。

（3）DISTRIBUTE BY 根据指定的内容将输入数据分给同一个 Reducer 操作。

（4）CLUSTER BY 除了具有 DISTRIBUTE BY 的功能之外，还会对指定字段的输入数据进行排序。

SELECT 语句的操作实例说明如下。

```sql
//全表查询、指定表字段查询
select * from emp;
select empno,ename from emp;
//条件过滤
select * from emp where deptno=10;
select * from emp where empno>=7500;
select ename,sal from emp where sal between 800 and 1500;
```

```sql
select * from emp limit 4;
select ename,sal,comm from emp where ename in ('Eula','Zora');
select ename,sal,comm from emp where comm is null;
//查询部门编号为 10 的部门员工数
select count(*) from emp where deptno=10;
//查询最高工资、最低工资、工资总和及平均工资
select max(sal),min(sal),sum(sal),avg(sal) from emp;
//查询每个部门的平均工资
select deptno,avg(sal) from emp group by deptno;
//查询平均工资大于 2000 的部门
select avg(sal),deptno from emp group by deptno having avg(sal)>2000;
```

8. Hive 中函数的使用方法

字符串长度函数为 length，语法格式为 length(string A)，应用实例如下。

```sql
//返回字符串"abcdefg"的长度
select length('abcdefg');
```

字符串反转函数为 reverse，语法格式为 reverse(string A)，应用实例如下。

```sql
//返回字符串"abc"的反转结果
select reverse('abc');
```

字符串连接函数为 concat，语法格式为 concat(string A, string B,…)，应用实例如下。

```sql
//返回输入字符串连接后的结果，支持输入任意个字符串
select concat('abc','def','gh');
```

带分隔符的字符串连接函数为 concat_ws，语法格式为 concat_ws(string SEP, string A, string B,…)，SEP 表示各个字符串间的分隔符应用实例如下。

```sql
//返回输入字符串连接后的结果
select concat_ws(',','abc','def','gh');
```

字符串截取函数为 substr、substring，语法格式为 substr(string A, int start)、substring(string A, int start)，二者用法相同，应用实例如下。

```sql
//返回字符串"abcde"从位置 3 到结尾的字符串
select substr('abcde',3);
```

返回指定字符个数的字符串截取函数为 substr，语法格式为 substr(string A, int start, int len)，应用实例如下。

```sql
//返回字符串"abcde"从位置 3 开始，且长度为 2 的字符串
select substr('abcde',3,2);
```

将字符串中的小写字母转换为大写字母的函数为 upper，语法格式为 upper(string A)，应用实例如下。

```sql
//将字符串"abSEd"中的小写字母转换为大写字母，并返回字符串
select upper('abSEd');
```

将字符串中的大写字母转换为小写字母的函数为 lower，语法格式为 lower(string A)，应用实例如下。

```
//将字符串"abSEd"中的大写字母转换为小写字母,并返回字符串
select lower('abSEd');
```
　　删除字符串两侧空格的函数为 trim,语法格式为 trim(string A),应用实例如下。
```
//删除字符串两侧的空格
select trim(' abc ');
```
　　删除字符串左侧空格的函数为 ltrim,语法格式为 ltrim(string A),应用实例如下。
```
//删除字符串左侧的空格
select ltrim(' abc ');
```
　　删除字符串右侧空格的函数为 rtrim,语法格式为 rtrim(string A),应用实例如下。
```
//删除字符串右侧的空格
select rtrim(' abc ');
```
　　正则表达式替换函数为 regexp_replace,语法格式为 regexp_replace(string A, string B, C),即将字符串 A 中的符合正则表达式 B 的部分替换为 C,应用实例如下。
```
select regexp_replace('foobar', 'oo|ar', '');
```
　　正则表达式解析函数为 regexp_extract,语法格式为 regexp_extract(string subject, string pattern, int index),即将字符串 subject 按照正则表达式 pattern 的规则拆分,返回指定的 index 字符,应用实例如下。
```
select regexp_extract('foothebar', 'foo(.*?)(bar)', 1);
```
　　统一资源定位符(Uniform Resource Locator,URL)解析函数为 parse_url,语法格式为 parse_url(string urlString, string partToExtract [,string keyToExtract]),该函数返回 URL 中指定的部分。partToExtract 的有效值为 HOST、PATH、QUERY、REF、PROTOCOL、AUTHORITY、FILE 和 USERINFO,应用实例如下。
```
select parse_url('https://www.ryjiaoyu.com', 'HOST') ;
```
　　JavaScript 对象表示法(JavaScript Object Notation,JSON)解析函数为 get_json_object,语法格式为 get_json_object(string json_string, string path),用于解析 JSON 字符串 json_string,并返回 path 指定的内容;如果输入的 JSON 字符串无效,那么返回 null,应用实例如下。
```
select get_json_object('{"store":
> {"fruit":\[{"weight":8,"type":"apple"},{"weight":9,"type":"pear"}],
> "bicycle":{"price":19.95,"color":"red"}
> },
> "email":"amy@only_for_json_udf_test.net",
> "owner":"amy"
> }
> ','$.owner');
```
　　空格字符串函数为 space,语法格式为 space(int n),该函数返回长度为 n 的空格字符串,应用实例如下。
```
//返回长度为 10 的空格字符串
select space(10);
```
　　重复字符串函数为 repeat,语法格式为 repeat(string str, int n),该函数返回对 str 重复 n 次后的字符串,应用实例如下。

项目 2　Hive 环境搭建与基本操作

```
//返回对"abc"重复 5 次后的字符串
select repeat('abc',5);
```

将字符串的首字符转换为美国信息交换标准码（American Standard Code for Information Interchange，ASCII）的函数为 ascii，语法格式为 ascii(string str)，用于返回字符串 str 首字符的 ASCII，应用实例如下。

```
//返回"a"的ASCII
select ascii('abc');
```

左补足函数为 lpad，语法格式为 lpad(string str, int len, string pad)，用于将字符串 str 以字符串 pad 左补足到 len 位，应用实例如下。

```
select lpad('abc',10,'td');
```

右补足函数为 rpad，语法格式为 rpad(string str, int len, string pad)，用于将字符串 str 以字符串 pad 右补足到 len 位，应用实例如下。

```
select rpad('abc',10,'td');
```

分割字符串函数为 split，语法格式为 split(string str, string pat)，用于按照字符串 pat 分割字符串 str，并返回分割后的字符串数组，应用实例如下。

```
select split('abtcdtef','t');
```

字符串集合查找函数为 find_in_set，语法格式为 find_in_set(string str, string strList)，用于返回字符串 str 在 strList 中第一次出现的位置，strList 是用逗号分隔的字符串集合，如果没有找到字符串 str，则返回 0，应用实例如下。

```
select find_in_set('ab','ef,ab,de');
```

9. 分区表操作

在 Hive 中，SELECT 查询一般会扫描整张表的数据，该操作将会导致查询性能的下降，同时，大部分查询操作实际上只需要扫描表中的部分数据，因此，为了解决这个问题，Hive 在创建表时引入了分区概念。Hive 的分区表指的是在创建表时指定分区空间。

Hive 可以将数据按照某列或者某些列作为一个分区进行分区管理，目前互联网应用每天都要存储大量的日志文件，其中必然会产生名为"日志日期"的列属性，因此，在分区时，可以按照日志产生的日期进行划分，把每天的日志当作一个分区。

Hive 将数据组织成分区，主要是为了提高数据的查询速度。而存储的每一条记录到底存储到哪个分区中是由用户决定的，即用户在加载数据的时候必须显式地指定该部分数据的存储分区。

假设在 Hive 创建的表中存在 id、content、d_date、d_time 这 4 列，则创建分区表的操作如下。

在表定义时创建单分区表，按照数据产生的日期属性进行分区，应用实例如下。

```
create table day_table (id int, content string,d_time string) partitioned by (d_date string);
```

在表定义时创建双分区表，按照数据产生的日期和时间属性进行分区，应用实例如下。

```
create table day_hour_table (id int, content string) partitioned by (d_date string, d_time string);
```

如果表已创建，则可以在此基础上添加分区，语法格式如下。

```
ALTER TABLE table_name ADD partition_spec [ LOCATION 'location1' ]
partition_spec [ LOCATION 'location2' ] ...;
```

如果分区已经存在，则可以对分区进行删除操作，语法格式如下。
```
ALTER TABLE table_name DROP partition_spec, partition_spec,...;
```
将数据加载到分区表中的语法格式如下。
```
LOAD DATA [LOCAL] INPATH 'filepath' [OVERWRITE] INTO TABLE
tablename [PARTITION (partcol1=val1, partcol2=val2 ...)];
```
查看分区操作的语法格式如下。
```
SHOW PARTITIONS table_name;
```

10. 分桶操作

Hive 会根据列的哈希值进行桶的组织。因此，对于每一张表或者每个分区，Hive 都可以将它们进一步组织成桶。如果说分区表操作是数据的粗粒度划分，而分桶操作是数据的细粒度划分。当数据量较大时，为提高数据的分析与计算速度，必须采用多个 Map 和 Reduce 任务来加快数据的处理速度。但是如果输入文件只有一个，则只能启动一个 Map 任务。此时，对数据进行分桶操作将会是一个很好的选择，即通过指定 clustered 字段计算列的哈希值，将文件分割成多个小文件，从而增加 Map 任务，以提高数据的分析与计算速度。

把表（或者分区）组织成桶的操作如下。
```
create table t_buck(id int, name string)clustered by(id)
sorted by(id)into 4 buckets
row format delimited
fields terminated by ','
TBLPROPERTIES("skip.header.line.count"="1");
```

将数据导入分桶表有如下两种方式。

（1）将外部生成的数据导入桶。这种方式不会自动分桶，如果导入数据的表没有进行分桶操作，则需要对导入数据的表提前进行分桶操作。其语法格式如下。
```
LOAD DATA LOCAL INPATH '/home/hadoop/user.dat' INTO TABLE t_user_buck;
```

（2）向已创建的分桶表中插入数据。此时插入数据的表应当为已分桶且排序的表。执行 INSERT 语句前需要设置 set hive.enforce.bucketing = true，以强制采用多个 Reduce 任务进行输出。
```
insert overwrite table t_buck select id,name from t_buck_from cluster by(id);
```

需要注意，创建分桶并不意味着导入操作导入的数据也是分桶的，因此，必须先对表（或者分区）进行分桶操作，再进行导入操作。

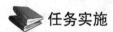 任务实施

Hadoop 和 Hive 的环境已经搭建好，下面进行查询测试。为了清楚地对比 MySQL 与 Hive 在不同数据量下的性能，本测试分为两部分进行。

1. 测试数据量较小时

测试数据量设定为 50 000 余行（测试中可以根据实际情况进行设定），具体测试过程如下。

微课 2-2　Hive 的应用

项目 2　Hive 环境搭建与基本操作

（1）MySQL 数据查询与提取

① 在测试用表中查询数据库的操作命令及结果如下。其中，测试用表对应的数据库为 info_station，数据表为 info_small01。

```
mysql> show databases;
+--------------------+
| Database           |
+--------------------+
| information_schema |
| info_station       |
| information_DB     |
| myhive             |
| mysql              |
| performance_schema |
| sys                |
+--------------------+
7 rows in set (0.00 sec)
```

② 在 MySQL 中执行 "select * from info_small01" 命令查询数据，进行测试，操作命令及部分结果如下。

```
mysql>select * from info_small01
| 115.171.192.42  | 2015/3/30 23:59 | /uc_server/images/noavatar_middle.gif | 8 |
| 180.173.113.181 | 2015/3/30 23:59 | /data/cache/style_1_forum_viewthread.
css?F97| 8 |
+-----------------+-----------------+---------------------------------------
--------------------------+------+
50368 rows in set (0.12 sec)
```

③ 将表 info_small01 中的数据导出到文件中（文件扩展名为 .csv），导出目录 "mysql-files" 需要提前创建，操作命令及结果如下。

```
mysql> select * from info_small01 into outfile '/var/lib/mysql-files/info_small01.
csv' fields terminated by ',' optionally enclosed by '"' escaped by '"' lines
terminated by '\r\n';
Query OK, 50368 rows affected (0.04 sec)
```

④ 在指定的导出目录下，查看生成的数据文件，操作命令及结果如下。

```
[root@master1 mysql-files]# ll
total 3980
-rw-rw-rw-. 1 mysql mysql 4073268 Mar 14 20:45 info_small01.csv
```

（2）向 Hive 中导入数据并进行查询

① 在 Hive 中创建表 info_small02，操作命令及结果如下。

```
hive> create table info_small02(
    > host string,
    > time string,
```

```
       > url string,
       > id int
       > ) row format delimited fields terminated by ',';
OK
Time taken: 0.628 seconds
```

② 导入数据的过程如下。将数据文件 info_small01.csv 导入 Hive 的表 info_small02 中，操作命令及结果如下。

```
hive> LOAD DATA LOCAL INPATH '/var/lib/mysql-files/info_small01.csv' INTO TABLE info_small02;
Loading data to table default.info_small02
Table default.info_small02 stats: [numFiles=1, totalSize=4073268]
OK
Time taken: 0.324 seconds
```

当数据被加载到表中时，不会对数据进行任何转换。导入操作只是将数据复制到 Hive 表对应的位置，这个表只有一个文件，文件没有被分割成多份。

在 HDFS 中查看生成的文件，操作命令及结果如下。

```
hive> dfs -ls /user/hive/warehouse/info_small02;
Found 1 items
-rwxr-xr-x   2 root supergroup    4073268 2023-03-14 20:56 /user/hive/warehouse/info_small02/info_small01.csv
```

③ 执行查询的过程如下。使用 Hive 对 50000 余行数据的文件执行简单查询操作，由操作结果可见耗费 0.271s，操作命令及结果如下。

```
hive> SELECT * FROM info_small02;
"115.171.192.42"       "2015/3/30 23:59"
"/uc_server/images/noavatar_middle.gif"  8
"60.164.35.90"  "2015/3/30 23:59"
"/home.php?mod=spacecp&ac=blog"  8
"115.171.192.42"       "2015/3/30 23:59"
"/uc_server/images/noavatar_middle.gif"  8
"115.171.192.42"       "2015/3/30 23:59"
"/uc_server/data/avatar/000/06/54/56_avatar_middle.jpg"  8
"115.171.192.42"       "2015/3/30 23:59"
"/uc_server/images/noavatar_middle.gif"  8
"180.173.113.181"      "2015/3/30 23:59"
"/data/cache/style_1_forum_viewthread.css?F97"  8
Time taken: 0.271 seconds, Fetched: 50368 row(s)
```

（3）优化导入过程

在优化导入过程中，通过指定 clustered 字段将文件通过哈希算法分割成多个小文件，装入不同的桶中。这里设置了 8 个桶。它们会为数据提供额外的结构以获得更快的查询速度。

项目 2　Hive 环境搭建与基本操作

① 新建表 info_small03，操作命令及结果如下。

```
hive> CREATE TABLE info_small03(
    > host string,
    > time string,
    > url string)
    > partitioned by (id int)
    > clustered by (time) sorted by (time) into 8 buckets
    > row format delimited fields terminated by ',';
OK
Time taken: 0.075 seconds
```

② 将数据从表 info_small02 中导入表 info_small03 中。

强制执行装桶的操作，操作命令如下。

```
hive>SET hive.enforce.bucketing=true;
```

导入数据，操作命令如下。

```
hive> FROM info_small02 insert overwrite table info_small03 partition(id=1) SELECT host,time,url WHERE id=1;
```

将表 info_small02 中的数据导入表 info_small03 中的部分结果如下。

```
2023-03-14 21:40:30,461 Stage-1 map = 0%,  reduce = 0%
2023-03-14 21:40:34,534 Stage-1 map = 100%, reduce = 0%, Cumulative CPU 1.51 sec
2023-03-14 21:40:39,717 Stage-1 map = 100%, reduce = 25%, Cumulative CPU 3.94 sec
2023-03-14 21:40:42,792 Stage-1 map = 100%, reduce = 38%, Cumulative CPU 5.29 sec
2023-03-14 21:40:43,820 Stage-1 map = 100%, reduce = 50%, Cumulative CPU 6.86 sec
2023-03-14 21:40:44,839 Stage-1 map = 100%, reduce = 100%, Cumulative CPU 12.68 sec
MapReduce Total cumulative CPU time: 12 seconds 680 msec
Ended Job = job_1678799927905_0016
Loading data to table default.info_small03 partition (id=7)
Partition default.info_small03{id=7} stats: [numFiles=8, numRows=6297, totalSize=490630, rawDataSize=484333]
MapReduce Jobs Launched:
Stage-Stage-1: Map: 1  Reduce: 8   Cumulative CPU: 12.68 sec   HDFS Read: 4106990 HDFS Write: 491249 SUCCESS
Total MapReduce CPU Time Spent: 12 seconds 680 msec
OK
Time taken: 19.729 seconds
```

③ 查看文件是否被分桶。从 Hadoop 集群的 HDFS Web 界面中查看所创建的文件分区，如图 2-3 所示。在 Web 界面中查看所创建的文件分桶，如图 2-4 所示。

在命令行中查看文件分桶，操作命令及结果如下。

```
hive> dfs -ls /user/hive/warehouse/info_small03/id=1;
Found 8 items
```

```
-rwxr-xr-x   2 root supergroup      130177 2023-03-14 21:35 /user/hive/warehouse/
info_small03/id=1/000000_0
-rwxr-xr-x   2 root supergroup      154686 2023-03-14 21:35 /user/hive/warehouse/
info_small03/id=1/000001_0
-rwxr-xr-x   2 root supergroup       28581 2023-03-14 21:35 /user/hive/warehouse/
info_small03/id=1/000002_0
-rwxr-xr-x   2 root supergroup           0 2023-03-14 21:35 /user/hive/warehouse/
info_small03/id=1/000003_0
-rwxr-xr-x   2 root supergroup          91 2023-03-14 21:35 /user/hive/warehouse/
info_small03/id=1/000004_0
-rwxr-xr-x   2 root supergroup           0 2023-03-14 21:35 /user/hive/warehouse/
info_small03/id=1/000005_0
-rwxr-xr-x   2 root supergroup       20674 2023-03-14 21:35 /user/hive/warehouse/
info_small03/id=1/000006_0
-rwxr-xr-x   2 root supergroup      154821 2023-03-14 21:35 /user/hive/warehouse/
info_small03/id=1/000007_0
```

Browse Directory

/user/hive/warehouse/info_small03

Permission	Owner	Group	Size	Last Modified	Replication	Block Size	Name
drwxr-xr-x	root	supergroup	0 B	2023/3/14 21:35:11	0	0 B	id=1
drwxr-xr-x	root	supergroup	0 B	2023/3/14 21:36:58	0	0 B	id=2
drwxr-xr-x	root	supergroup	0 B	2023/3/14 21:38:29	0	0 B	id=3
drwxr-xr-x	root	supergroup	0 B	2023/3/14 21:38:56	0	0 B	id=4
drwxr-xr-x	root	supergroup	0 B	2023/3/14 21:39:23	0	0 B	id=5
drwxr-xr-x	root	supergroup	0 B	2023/3/14 21:39:53	0	0 B	id=6
drwxr-xr-x	root	supergroup	0 B	2023/3/14 21:40:46	0	0 B	id=7
drwxr-xr-x	root	supergroup	0 B	2023/3/14 21:50:39	0	0 B	id=8

图 2-3　查看所创建的文件分区

Browse Directory

/user/hive/warehouse/info_small03/id=1

Permission	Owner	Group	Size	Last Modified	Replication	Block Size	Name
-rwxr-xr-x	root	supergroup	127.13 KB	2023/3/14 21:35:04	2	128 MB	000000_0
-rwxr-xr-x	root	supergroup	151.06 KB	2023/3/14 21:35:03	2	128 MB	000001_0
-rwxr-xr-x	root	supergroup	27.91 KB	2023/3/14 21:35:06	2	128 MB	000002_0
-rwxr-xr-x	root	supergroup	0 B	2023/3/14 21:35:07	2	128 MB	000003_0
-rwxr-xr-x	root	supergroup	91 B	2023/3/14 21:35:08	2	128 MB	000004_0
-rwxr-xr-x	root	supergroup	0 B	2023/3/14 21:35:08	2	128 MB	000005_0
-rwxr-xr-x	root	supergroup	20.19 KB	2023/3/14 21:35:09	2	128 MB	000006_0
-rwxr-xr-x	root	supergroup	151.19 KB	2023/3/14 21:35:09	2	128 MB	000007_0

图 2-4　查看所创建的文件分桶

④ 执行查询的过程如下。在 Hive 中执行 "select * from info_small03" 命令,操作命令及部分结果如下。

```
hive>SELECT * FROM info_small03
"111.194.234.168"      "2015/3/30 23:55"
"/home.php?mod=spacecp&ac=follow&op=checkfeed&rand=1369929316"    8
"118.207.155.181"      "2015/3/30 23:55"
"/uc_server/avatar.php?uid=65434&size=small"    8
"111.194.234.168"      "2015/3/30 23:55"
"/static/image/common/vlineb.png"    8
"222.36.188.206"       "2015/3/30 23:55"
"/static/image/common/tip_bottom.png"    8
"223.208.90.114"       "2015/3/30 23:55"
"/member.php?mod=logging&action=login"    8
"222.36.188.206"       "2015/3/30 23:55"
"/static/image/common/tip_bottom.png"    8
Time taken: 0.038 seconds, Fetched: 50367 row(s)
```

在前面的操作中,使用的测试数据为 50 000 余行。由测试结果可以看出 MySQL 查询 50 000 行左右的数据大概需要 0.04s,用优化前的 Hive 对相同规模的数据进行查询大概需要 0.271s,用优化后的 Hive 进行同样的查询大概需要 0.038s。由此可见,在小数据量的情况下,完成同样的查询操作,Hive 与 MySQL 相比并没有体现出性能上的优势。下面采用大数据量(约 7000 万行)进行操作对比。

2. 测试数据量较大时

测试数据量设定为约 7000 万行(测试中可以根据实际进行设定,数据量越大越好),具体测试过程与小数据量时的测试过程类似,测试结果如下。

(1) MySQL 查询结果

启动 MySQL,创建数据表 info1 并导入数据,查询 7000 多万行数据,结果直接显示 "Killed",操作命令及结果如下。

```
#创建数据表 info1
mysql>CREATE TABLE `info1` (
 >`zona` varchar(255),
 >`local` varchar(255),
 >`secao` varchar(255),
 >`qtd_eleitores_biometrico` varchar(255),
 >`qtd_eleitores_codigo` varchar(255),
 >`id_eleicao` varchar(255),
 >`qtd_eleitores_aptos` varchar(255),
 >`tipo_cargo` varchar(255),
 >`codigo_cargo` varchar(255),
 >`qtd_comparecimento` varchar(255),
```

```
>`tipo_voto` varchar(255),
>`qtd_votos` varchar(255),
>`partido` varchar(255),
>`codigo_candidato` varchar(255),
>`identificacao_votavel` varchar(255)
>) ENGINE=InnoDB DEFAULT CHARSET=utf8mb3;

#导入数据
mysql>LOAD DATA LOCAL INFILE '/opt/info.csv' INTO TABLE info1 FIELDS TERMINATED BY ',' IGNORE 1 LINES;
Query OK, 70115756 rows affected, 65535 warnings (5 min 31.57 sec)
Records: 70115756  Deleted: 0  Skipped: 0  Warnings: 65477431

#查询全表数据
mysql> SELECT * FROM info1;
Killed
```

（2）Hive 优化前的查询结果

进入 Hive 命令行，创建数据表 info2，表结构如下。

```
hive>CREATE TABLE info2(
>zona string,
>local_x string,
>secao string,
>qtd_eleitores_biometrico string,
>qtd_eleitores_codigo string,
>id_eleicao string,
>qtd_eleitores_aptos string,
>tipo_cargo string,
>codigo_cargo string,
>qtd_comparecimento string,
>tipo_voto string,
>qtd_votos string,
>partido string,
>codigo_candidato string,
>identificacao_votavel string)
>row format delimited fields terminated by ','
>TBLPROPERTIES("skip.header.line.count"="1");
```

将数据导入数据表 info2，操作命令及结果如下。

```
hive>LOAD DATA LOCAL INPATH '/opt/info.csv' OVERWRITE INTO TABLE info2;
Loading data to table default.info2
Table default.info2 stats: [numFiles=1, numRows=0, totalSize=7573040744, rawDataSize=0]
```

```
OK
Time taken: 27.494 seconds
```
执行查询操作，操作命令及结果如下。
```
hive>SELECT * FROM info2;
Query ID = root_20230330122426_3b5b542a-47d3-4f75-955f-253440c2d3fc
Total jobs = 1
Launching Job 1 out of 1
Number of reduce tasks is set to 0 since there's no reduce operator
Starting Job = job_1680148389853_0002, Tracking URL = http://master1:8088/
proxy/application_1680148389853_0002/
Kill Command = /usr/local/hadoop/bin/hadoop job  -kill job_1680148389853_0002
Hadoop job information for Stage-1: number of mappers: 39; number of reducers: 0
2023-03-30 12:24:31,792 Stage-1 map = 0%,  reduce = 0%
……
41   1023    125 55  17  546 155 majoritario governador   121 nulo      3    -   -   -
41   1023    125 55  17  544 155 majoritario presidente   121 nominal 2    12  12
     {'partido': 12
41   1023    125 55  17  544 155 majoritario presidente   121 branco    3    -   -   -

Time taken: 86.859 seconds, Fetched: 70115756 row(s)
```
执行筛选查询操作，操作命令及结果如下。
```
hive>SELECT * FROM WHERE zona ="1"
……
1    2208    719 126 3   544 274 majoritario presidente   239 nominal 10   12  12
     {'partido': 12
1    2208    719 126 3   544 274 majoritario presidente   239 nominal 124  13  13
     {'partido': 13
1    2208    719 126 3   544 274 majoritario presidente   239 nulo      1    -   -   -

Time taken: 57.607 seconds, Fetched: 1380590 row(s)
```
（3）Hive 优化后的查询结果

进入 Hive 命令行，创建数据表 info3，表结构如下。
```
#创建数据表 info3
hive>CREATE TABLE info3(
>local_x string,
>secao string,
>qtd_eleitores_biometrico string,
>qtd_eleitores_codigo string,
>id_eleicao string,
>qtd_eleitores_aptos string,
```

```
>tipo_cargo string,
>codigo_cargo string,
>qtd_comparecimento string,
>tipo_voto string,
>qtd_votos string,
>partido string,
>codigo_candidato string,
>identificacao_votavel string)
>PARTITIONED BY(zona string)
>CLUSTERED BY(local_x)SORTED BY(local_x) INTO 8 BUCKETS
>ROW FORMAT DELIMITED FIELDS TERMINATED BY ',';
```

设置强制分桶，操作命令如下。

```
hive>SET hive.enforce.bucketing=true
```

修改指定分区内的数据，操作命令及结果如下。

```
hive>INSERT OVERWRITE TABLE info3
>partition(zona="1")
>select
local_x,secao,qtd_eleitores_biometrico,qtd_eleitores_codigo,id_eleicao,qtd_
eleitores_aptos,tipo_cargo,codigo_cargo,qtd_comparecimento,tipo_voto,qtd_votos,
partido,codigo_candidato,identificacao_votavel from info2 where zona = "1";

#查询分区
hive>SELECT * FROM info3 WHERE zona ='1';
......
1473    212 26  8    546 35  proporcional    deputadoEstadual    30  nominal 1   22
22222   {'partido': 22 1
1473    212 26  8    546 35  proporcional    deputadoEstadual    30  nominal 2   22
22000   {'partido': 22 1

Time taken: 0.036 seconds, Fetched: 1380590 row(s)
```

由前面的测试可以看出，在大数据量的情况下，利用 Hive 与 MySQL 完成同样的查询操作，Hive 性能优势比较明显。Hive 对以 GB 为单位量级的数据增长是不敏感的，而当 MySQL 的数据量增长到约 7GB 时，要完成查询需要通过各种优化程序对查询过程进行优化处理。

项目小结

本项目介绍了 Hive 产生的背景、Hive 的架构、特点和执行流程，完成了 Hive 的安装与配置等操作；并重点介绍了 Hive SQL 基本操作，包括创建内部表、创建外部表、修改表以及查看数据库和表的相关信息、复制和移动数据操作、分区表和分桶的创建方法等。通过本

项目 2　Hive 环境搭建与基本操作

项目的学习，读者能够对 Hive 的相关技术与应用建立清晰的认识，为学习大数据存储的相关知识和技能奠定良好的基础。

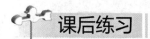

1. Hive 内部表和外部表有哪些区别？
2. Hive 提供哪些方式保存元数据，它们各有什么特点？

项目 ❸ ZooKeeper 环境搭建与应用

学习目标

【知识目标】
了解 ZooKeeper 的功能。
识记 ZooKeeper 各组件的功能与联系。

【技能目标】
学会 ZooKeeper 的安装与配置。
学会 ZooKeeper 节点管理的相关操作。

【素质目标】
具有严谨细致的工作态度和工作作风。
具有良好的团队协作意识和业务沟通能力。
具有良好的表达能力和文档查阅能力。
具有规范的编程意识和较好的数据洞察力。

项目描述

ZooKeeper 主要用来解决分布式集群中应用系统一致性的问题,它能提供基于类似文件系统的目录节点树方式的数据存储方式。ZooKeeper 用来维护和监控存储数据的状态变化,通过监控这些数据状态的变化,实现基于数据的集群管理。

本项目主要引导学生完成 ZooKeeper 的安装与配置,并学习 ZooKeeper 节点管理的相关操作。

任务 3.1　ZooKeeper 的安装与配置

 任务描述

(1)学习 ZooKeeper 的相关知识,了解 ZooKeeper 在分布式管理上的优势。
(2)完成 ZooKeeper 在 3 种模式下的安装与配置。

任务目标

(1)熟悉 ZooKeeper 的功能。
(2)学会 ZooKeeper 的安装与配置。

项目 3　ZooKeeper 环境搭建与应用

　知识准备

ZooKeeper 是一个开源的分布式协调服务，由雅虎公司创建，是谷歌 Chubby 的开源实现。分布式应用可以基于 ZooKeeper 实现诸如数据发布/订阅、负载均衡、命名服务、分布式协调/通知、集群管理、领导者（Leader）选举、分布式锁和分布式队列等功能。在分布式环境中，协调和管理服务是一个复杂的过程。ZooKeeper 通过其简单的架构和 API 解决了这个问题。ZooKeeper 允许开发人员专注于核心应用程序的逻辑，而不必担心应用程序的分布式特性。

1. 分布式应用

在进一步学习 ZooKeeper 之前，先了解一下分布式应用的优点和面临的挑战，这有利于理解 ZooKeeper 的优势。

在给定时间内，分布式应用可以同时在网络中的多个系统上运行，通过协调这些系统，分布式应用可以快速、有效地完成特定任务。通常来说，对于复杂而耗时的任务，非分布式应用（运行在单个系统中）需要较长时间才能完成，而分布式应用通过使用多个系统的计算能力可以在短时间内完成。

通过将分布式应用配置在多个系统上运行，可以进一步缩短完成任务的时间。运行分布式应用的一组机器称为集群，而在集群中运行的每台机器都被称为节点。

在分布式应用工作过程中，主要涉及服务器和客户端两类设备。服务器实际上是分布式的，并具有通用接口，以便客户端连接到集群中的任意服务器上并获得相同的结果；客户端是与分布式应用进行交互的工具。ZooKeeper 应用范例如图 3-1 所示。

分布式应用具有很多优点。分布式应用的容错能力较强，诸如单个或几个系统的故障不会使整个系统出现故障。分布式应用在性能扩展上也具备较高的灵活性，它可以通过添加更多机器来提高系统性能，而无非中断服务。分布式应用可以隐藏系统的复杂性，并将其显示为单个实体/应用。

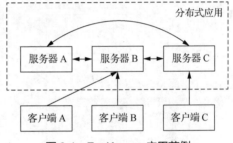

图 3-1　ZooKeeper 应用范例

同时，分布式应用也面临着诸多挑战，例如，在两个或多个机器尝试执行特定任务时，共享资源只能在任意给定时间内由单台机器修改而引发的竞争条件设定问题；两个或多个操作无限期等待彼此完成操作而引发的死锁问题；数据操作的部分失败而引发的数据的不一致问题。

综上，分布式应用提供了很多好处，但它也抛出了一些复杂和难以解决的挑战。ZooKeeper 框架提供了一个完整的机制来应对这些挑战。对于竞争条件设定和死锁问题，ZooKeeper 通过安全同步方法进行规避；对于数据的不一致问题，ZooKeeper 使用原子性解析来应对。ZooKeeper 是由集群（节点组）使用的一种服务，用于在自身之间进行协调，并通过稳健的同步技术维护共享数据。ZooKeeper 本身是一个分布式协调服务，为写入分布式应用提供服务。ZooKeeper 提供的常见服务如下。

（1）命名

该服务按名称标识集群中的节点。它类似于域名系统（Domain Name System，DNS），但仅应用于节点。

（2）配置管理

该服务可以为节点添加最新的系统配置信息。

（3）集群管理

该服务实时地在集群中感知节点状态，如节点的死亡和新节点的加入。

（4）选举算法

该服务选举一个节点作为达成协调目的的 Leader。

（5）锁定和同步

该服务在修改数据的同时锁定数据，可帮助用户在连接其他分布式应用（如 Apache HBase）时进行自动故障恢复。

（6）高度可靠的数据注册表

该服务能保证在一个或几个节点关闭时仍然可以获得数据。

2. ZooKeeper 的基本概念

（1）ZooKeeper 的架构

ZooKeeper 的架构如图 3-2 所示。

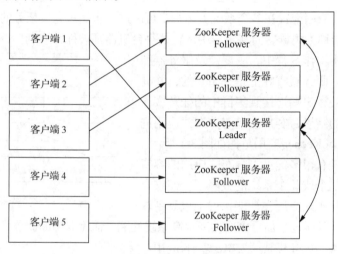

图 3-2　ZooKeeper 的架构

ZooKeeper 中的组件如表 3-1 所示。

表 3-1　ZooKeeper 中的组件

组件	描述
客户端	客户端是 ZooKeeper 中的一个节点，从服务器访问信息。经过特定的时间间隔，每个客户端都向服务器发送消息以使服务器知道客户端是活跃的。 类似地，当客户端连接时，会向服务器发送确认码。如果连接的服务器没有响应，则客户端会自动将消息重定向到另一个服务器
服务器	服务器是 ZooKeeper 中的一个节点，为客户端提供所有的服务，向客户端发送确认码以告知服务器是活跃的
服务器组（Ensemble）	服务器组由多个节点组成，ZooKeeper 形成服务器组所需的最小节点数为 3
领导者（Leader）	对于服务器节点，如果其任意连接节点失败，则执行自动恢复功能。Leader 在服务启动时被选举
跟随者（Follower）	Follower 是跟随 Leader 指令的服务器节点

（2）层次命名空间

图 3-3 描述了用于内存表示的 ZooKeeper 文件系统的树结构，即 ZooKeeper 数据结构。ZooKeeper 节点称为 znode。每个 znode 由一个名称标识，并用路径（/）分隔。

在图 3-3 中，有一个名称标识为"/"的 znode，其下有两个逻辑命名空间，即/config 和/workers。其中，/config 命名空间用于集中式配置管理，/workers 命名空间用于命名。在/config 命名空间下，每个 znode 最多可存储 1MB 的数据。ZooKeeper 数据结构的主要目的是存储同步数据并描述 znode 的元数据。

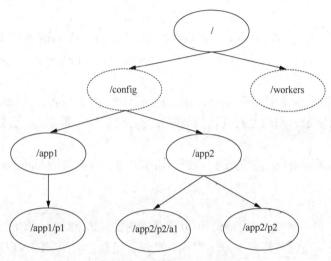

图 3-3 ZooKeeper 数据结构

ZooKeeper 数据结构中的每个 znode 都维护着一个 stat 结构。一个 stat 结构仅提供一个 znode 的元数据。它由版本号、操作控制列表、时间戳和数据长度组成。

① 版本号。每个 znode 都有版本号，这意味着当与 znode 相关联的数据发生变化时，其对应的版本号也会增加。当多个 ZooKeeper 客户端尝试在同一 znode 上执行操作时，通过比较和验证版本号，ZooKeeper 能够确保操作的有序性和一致性。

② 操作控制列表。其基本上是访问 znode 的认证机制，管理着所有 znode 的读取和写入操作。

③ 时间戳。时间戳表示创建和修改 znode 所经过的时间。它通常以毫秒为单位。ZooKeeper 以"事务 ID（zxid）"标识 znode 的每次更改。zxid 是唯一的，并且为每个事务保留时间，以便可以轻松地确定从一个请求到另一个请求所经过的时间。

④ 数据长度。存储在 znode 中的数据总量是数据长度，在 znode 中最多可以存储 1MB 的数据。

（3）znode 的类型

znode 被分为持久节点、临时节点和顺序节点。

① 持久节点。即使在创建该特定 znode 的客户端与 ZooKeeper 断开连接后，持久节点仍然存在。默认情况下，所有 znode 都是持久节点。

② 临时节点。客户端活跃时，临时节点就是有效的。当客户端与 ZooKeeper 集群断开连

接时,ZooKeeper 会自动删除临时节点。因此,临时节点不允许有子节点。如果临时节点被删除,则下一个合适的节点将填充其位置。临时节点是 Leader 选举过程的关键节点。

③ 顺序节点。顺序节点可以是持久的或临时的。当一个新的 znode 被创建为顺序节点时,ZooKeeper 会把一个 10 位的数字序列号附加到该顺序节点原始路径名称后作为其新的路径。例如,将路径为/myapp 的 znode 创建为顺序节点时,ZooKeeper 会将路径更改为/myapp0000000001,并将下一个序列号设置为 0000000002。如果两个顺序节点是同时创建的,那么 ZooKeeper 将对每个顺序节点使用不同的数字序列号。顺序节点是锁定和同步过程的关键节点。

(4) 会话

会话(Session)对于 ZooKeeper 的操作非常重要。会话中的请求按先进先出(First In First Out,FIFO)顺序执行。一旦客户端连接到服务器,ZooKeeper 就会建立会话并向客户端分配会话 ID。

客户端以特定的时间间隔向服务器发送心跳信号以保持会话有效。如果 ZooKeeper 集群在服务器开启时设定的时间内没有从客户端接收到心跳信号,即意味着会话超时,则它会判定客户端死机。

会话超时的时长通常以毫秒为单位。当会话结束时,在该会话期间创建的临时节点也会被删除。

(5) 监视

监视(Watch)是一种简单的机制,可以使客户端收到 ZooKeeper 集群中关于 znode 更改的通知。客户端可以在读取特定 znode 时设置监视。监视会向注册的客户端发送任何 znode 更改的通知。

znode 更改是指与 znode 相关的数据修改,或者 znode 子项的数据修改,其只触发一次监视。如果客户端想要再次收到监视发送的 ZooKeeper 集群中关于 znode 更改的通知,则必须通过另一个读取操作来完成。当会话连接过期时,客户端将与服务器断开连接,相关的监视也将被删除。

3. ZooKeeper 工作流

ZooKeeper 集群启动后,它将等待客户端连接到 ZooKeeper 集群中的一个节点,这个节点可以是 Leader 或 Follower。客户端连接成功后,节点将向特定客户端分配会话 ID,并向该客户端发送确认信号。如果客户端没有收到确认信号,则尝试连接 ZooKeeper 集群中的另一个节点。节点连接成功后,客户端以有规律的间隔向节点发送心跳信号,以确保连接有效。

如果客户端想要读取特定的 znode,则其会向具有 znode 路径的节点发送读取请求,相关节点会从自己的数据库中查询 znode 路径,并将路径信息发送给请求的客户端。因此,在 ZooKeeper 集群中,读取速度很快。

如果客户端想要将数据存储在 ZooKeeper 集群中,则其会将 znode 路径和数据发送到服务器中由服务器将该请求转发给 Leader,Leader 将向所有的 Follower 重新发出写入请求。如果大部分(超过半数)节点成功响应,则写入请求成功,否则,写入请求失败。

ZooKeeper 工作流如图 3-4 所示。

项目 3 ZooKeeper 环境搭建与应用

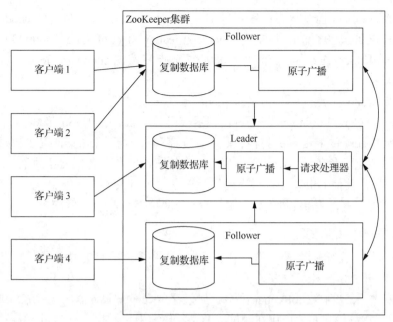

图 3-4 ZooKeeper 工作流

ZooKeeper 中各组件的功能如表 3-2 所示。

表 3-2 ZooKeeper 中各组件的功能

组件	功能
复制数据库（Replicated Database）	用于在 ZooKeeper 中存储数据。每个 znode 都有自己的数据库，由于数据一致性，各 znode 的数据相同
Leader	负责处理写入请求
Follower	负责从客户端接收写入请求，并将其转发到 Leader
请求处理器（Request Processor）	只存在于 Leader 中。它管理来自 Follower 的写入请求
原子广播（Atomic Broadcasts）	负责广播从 Leader 到 Follower 的变化

4. ZooKeeper Leader 选举

如何在 ZooKeeper 集群中选举 Leader 呢？考虑一个集群中有 N 个节点，Leader 选举的过程如下。

（1）所有节点创建具有相同路径 /app/leader_election/guid_ 的顺序、临时节点。

（2）ZooKeeper 集群将附加一个 10 位数字序列号到节点路径中，创建的 znode 的路径名称标识将是 /app/leader_election/guid_0000000001、/app/leader_election/guid_0000000002 等。

（3）对于给定的 ZooKeeper 实例，将 znode 中序列号的节点作为 Leader，而其他节点均为 Follower。

（4）每个 Follower 监视比自己序列号小一位的 znode。例如，创建的 znode/app/leader_election/guid_0000000008 节点将监视 znode/app/leader_election/guid_0000000007 节点，创建的 znode/app/leader_election/guid_0000000007 节点将监视 znode/app/leader_election/guid_0000000006 节点。

（5）如果 Leader 节点，则其相应的 znode 会被删除。

（6）监视 Leader 的 Follower 将通过监视器获得关于 Leader 删除的通知。

（7）监视 Leader 的 Follower 将检查是否存在序列号小于自己的 znode。如果不存在，其将承担 Leader 节点的角色；否则，则序列号更小的 znode 将承担 Leader 的角色。

（8）类似地，其他 Follower 选举序列号小于自己的 znode 作为 Leader。

Leader 选举是一个复杂的过程，但 ZooKeeper 服务使它变得非常简单。

5. ZooKeeper 的安装

ZooKeeper 的安装分为 3 种模式：单机模式、伪集群模式和集群模式。

（1）ZooKeeper 的单机模式安装

在这种模式下，没有 ZooKeeper 副本，如果 ZooKeeper 服务器出现故障，则 ZooKeeper 服务将会停止。这种模式主要应用在测试或 Demo 环境中，在实际生产环境中一般不会采用。

（2）ZooKeeper 的伪集群模式安装

伪集群模式就是在单机中模拟集群的 ZooKeeper 服务。在 ZooKeeper 的参数配置中，clientPort 参数用来配置客户端连接 ZooKeeper 的端口。伪集群模式使每个配置文档都模拟一台机器，也就是说，需要在单台机器中运行多个 ZooKeeper 实例，而且必须保证各个配置文档的 clientPort 参数不冲突。

（3）ZooKeeper 的集群模式安装

在这种模式下，可以获得可靠的 ZooKeeper 服务，只要集群中超过半数的 ZooKeeper 服务启动了，那么总的 ZooKeeper 服务就是可用的。集群模式与伪集群模式配置最大的不同是，集群模式中的 ZooKeeper 实例分布在多台机器上。

6. ZooKeeper 中的监视机制

ZooKeeper 的监视机制主要包括客户端线程、客户端 WatchManager 和 ZooKeeper 服务器 3 部分。其具体工作流程是，客户端线程在向 ZooKeeper 服务器注册监视的同时，会将监视对象存储在客户端 WatchManager 中；当 ZooKeeper 服务器触发监视事件后，会向客户端线程发送通知，客户端线程从客户端 WatchManager 中取出对应的监视对象来执行回调逻辑。

 任务实施

1. ZooKeeper 单机模式搭建

（1）安装所需软件

① 复制所需的安装包到容器中，操作命令及结果如下。

```
[root@CentOS opt]# docker cp zookeeper-3.4.8.tar.gz master1:/opt
Successfully copied 22.26MB to master1:/opt
```

微课 3-1
Zookeeper 的安装与配置 1

项目 3 ZooKeeper 环境搭建与应用

② 进入容器的安装包存储目录进行查看，操作命令及结果如下。

```
[root@master1 opt]# ll
total 1103808
-rw-r--r--. 1 root root  92834839 Feb 17 03:27 apache-hive-1.2.1-bin.tar.gz
-rw-r--r--. 1 root root 218720521 Feb  5 16:03 hadoop-2.7.7.tar.gz
-rw-r--r--. 1 root root 148162542 Feb  5 16:03 jdk-8u341-linux-x64.tar.gz
-rw-r--r--. 1 root root 644399365 Feb 15 15:55 mysql-5.7.23-linux-glibc2.12-
x86_64.tar.gz
drwxr-xr-x. 4 root root       151 Sep 24  2016 mysql-connector-java-5.1.40
-rw-r--r--. 1 root root   3911557 Feb 15 15:55 mysql-connector-java-5.1.40.tar.gz
-rw-r--r--. 1 root root  22261552 Feb 15 15:56 zookeeper-3.4.8.tar.gz
```

③ 在安装包存储目录中执行"tar -zxvf zookeeper-3.4.8.tar.gz -C /usr/local/zookeeper/"命令，将准备好的 ZooKeeper 安装包解压到指定目录中。进入解压路径并执行"mv zookeeper-3.4.8 zookeeper"命令，修改 ZooKeeper 的文件名（以方便后面的操作），操作命令如下。

```
[root@master1 opt]# tar -zxvf zookeeper-3.4.8.tar.gz -C /usr/local/zookeeper/
[root@master1 opt]# cd /usr/local/zookeeper/
[root@master1 zookeeper]# mv zookeeper-3.4.8 zookeeper
```

④ 进入 zookeeper/conf 目录，查看当前目录中的文件，执行"cp zoo_sample.cfg zoo.cfg"命令，复制 zoo_sample.cfg 文件并将其重命名为 zoo.cfg，再次查看当前目录中的文件，操作命令及结果如下。

```
[root@master1 zookeeper]# cd conf
[root@master1 conf]# ls
configuration.xsl  log4j.properties  zoo_sample.cfg
[root@master1 conf]# cp zoo_sample.cfg zoo.cfg
[root@master1 conf]# ls
configuration.xsl  log4j.properties  zoo.cfg  zoo_sample.cfg
```

（2）配置 zoo.cfg 文件

① 配置 zoo.cfg 文件，修改配置项如下。

```
tickTime=2000
#需要事先创建好 data 目录
dataDir=/usr/local/zookeeper/zookeeper/data
clientPort=2181
```

相关说明如下。

- tickTime。用于设定 ZooKeeper 服务器之间或客户端与服务器之间心跳信号的时间间隔。
- dataDir。用于设定 ZooKeeper 保存数据的目录，默认情况下，ZooKeeper 将写数据的日志文件也保存在这个目录中。
- clientPort。用于设定 ZooKeeper 服务器监视端口，用来接收客户端的访问请求。

② 配置好 zoo.cfg 文件后，即可启动 ZooKeeper 服务，进入 zookeeper/bin 目录，执行"./zkServer.sh start"命令，启动 ZooKeeper 服务。如果看到"Starting zookeeper ... STARTED"

返回信息，则表示 ZooKeeper 服务启动成功，操作命令及结果如下。

```
[root@master1 conf]# cd ..
[root@master1 zookeeper]# cd bin
[root@master1 bin]# ./zkServer.sh start
ZooKeeper JMX enabled by default
Using config: /usr/local/zookeeper/zookeeper/bin/../conf/zoo.cfg
Starting zookeeper ... STARTED
```

执行"./zkServer.sh status"命令，查看 ZooKeeper 服务的运行状态，操作命令及结果如下。

```
[root@master1 bin]# ./zkServer.sh status
ZooKeeper JMX enabled by default
Using config: /usr/local/zookeeper/zookeeper/bin/../conf/zoo.cfg
Mode: standalone
```

2. ZooKeeper 伪集群模式搭建

（1）安装所需软件

① 复制所需的安装包到容器中，操作命令及结果如下。

```
[root@CentOS opt]# docker cp zookeeper-3.4.8.tar.gz master1:/opt
Successfully copied 22.26MB to master1:/opt
```

② 进入容器的安装包所在目录进行查看，操作命令及结果如下。

```
[root@master1 opt]# ll
total 1103808
-rw-r--r--. 1 root root  92834839 Feb 17 03:27 apache-hive-1.2.1-bin.tar.gz
-rw-r--r--. 1 root root 218720521 Feb  5 16:03 hadoop-2.7.7.tar.gz
-rw-r--r--. 1 root root 148162542 Feb  5 16:03 jdk-8u341-linux-x64.tar.gz
-rw-r--r--. 1 root root 644399365 Feb 15 15:55 mysql-5.7.23-linux-glibc2.12-x86_64.tar.gz
drwxr-xr-x. 4 root root       151 Sep 24  2016 mysql-connector-java-5.1.40
-rw-r--r--. 1 root root   3911557 Feb 15 15:55 mysql-connector-java-5.1.40.tar.gz
-rw-r--r--. 1 root root  22261552 Feb 15 15:56 zookeeper-3.4.8.tar.gz
```

③ 在存储目录中执行"tar -zxvf zookeeper-3.4.8.tar.gz -C /usr/local/zookeeper/"命令，将准备好的 ZooKeeper 安装包解压到指定目录中。进入解压路径并执行"mv zookeeper-3.4.8 zookeeper"命令，修改 ZooKeeper 的文件名（以方便后面的操作），操作命令及结果如下。

```
[root@master1 opt]# tar -zxvf zookeeper-3.4.8.tar.gz -C /usr/local/zookeeper/
[root@master1 opt]# cd /usr/local/zookeeper/
[root@master1 zookeeper]# mv zookeeper-3.4.8 zookeeper
```

④ 进入 zookeeper/conf 目录，查看当前目录中的文件，执行"cp zoo_sample.cfg zoo.cfg"命令，复制 zoo_sample.cfg 文件并将其重命名为 zoo.cfg，再次查看当前目录中的文件，操作命令及结果如下。

```
[root@master1 zookeeper]# cd conf
[root@master1 conf]# ls
configuration.xsl  log4j.properties  zoo_sample.cfg
```

项目 3　ZooKeeper 环境搭建与应用

```
[root@master1 conf]# cp zoo_sample.cfg zoo.cfg
[root@master1 conf]# ls
configuration.xsl  log4j.properties  zoo.cfg  zoo_sample.cfg
```

（2）配置 zoo.cfg 文件

① 配置 zoo.cfg 文件，其中配置项"dataDir"和"dataLogDir"中的目录需要手动创建，具体配置内容如下。

```
tickTime=2000
initLimit=10
syncLimit=5
dataDir=/usr/local/zookeeper/zookeeper/data
dataLogDir=/usr/local/zookeeper/zookeeper/datalog
clientPort=2181
server.1=127.0.0.1:2888:3888
server.2=127.0.0.1:2889:3889
server.3=127.0.0.1:2890:3890
maxClientCnxns=60
```

相关说明如下。

- initLimit。该配置项表示允许 Follower 服务器连接并同步到 Leader 服务器的初始化时间，它以 tickTime 的倍数来表示。如果超过此时间，则连接失败。
- syncLimit。该配置项表示 Leader 服务器与 Follower 服务器之间信息同步允许的最大时间间隔，它以 tickTime 的倍数来表示。如果超过此时间，则 Follower 服务器与 Leader 服务器之间将断开连接。
- dataLogDir。用于设定保存 ZooKeeper 日志的路径。
- server.A=B:C:D。其中，A 是一个数字，代表这是第几号服务器；B 表示服务器的 IP 地址；C 表示服务器与集群中的 Leader 交换信息的端口；D 表示当 Leader 失效后，选举时服务器相互通信的端口。
- maxClientCnxns。用于设定连接到 ZooKeeper 服务器的客户端的最大数量。

② 在 ZooKeeper 安装目录中创建 data 和 datalog 目录，进入 data 目录并创建 myid 文件，将其内容编辑为 1，保存并退出，随后查看 data 目录中的文件，操作命令及结果如下。

```
[root@master1 zookeeper]# mkdir data
[root@master1 zookeeper]# mkdir datalog
[root@master1 zookeeper]# cd data
[root@master1 data]# vi myid
[root@master1 data]# ll
total 8
-rw-r--r--. 1 root root  2 Feb 18 01:44 myid
drwxr-xr-x. 2 root root 65 Feb 18 02:15 version-2
-rw-r--r--. 1 root root  4 Feb 18 02:14 zookeeper_server.pid
```

③ 进入/usr/local/zookeeper 目录，执行"cp －r zookeeper zookeeper1""cp －r zookeeper zookeeper2"命令，新建两个服务器，操作命令及结果如下。

```
[root@master1 zookeeper]# cp -r zookeeper zookeeper1
[root@master1 zookeeper]# cp -r zookeeper zookeeper2
[root@master1 zookeeper]# ls
zookeeper  zookeeper1  zookeeper2
```

④ 分别编辑新建的 zookeeper1 和 zookeeper2 中的 data/myid 文件,将 myid 文件的内容分别修改为 2 和 3。

分别编辑新建的 zookeeper1 和 zookeeper2 中的 conf/zoo.cfg 文件,zookeeper1 中的 zoo.cfg 文件内容如下。

```
tickTime=2000
initLimit=10
syncLimit=5
dataDir=/usr/local/zookeeper/zookeeper1/data
dataLogDir=/usr/local/zookeeper/zookeeper1/datalog
clientPort=2182
server.1=127.0.0.1:2888:3888
server.2=127.0.0.1:2889:3889
server.3=127.0.0.1:2890:3890
maxClientCnxns=60
```

zookeeper2 中的 zoo.cfg 文件内容如下。

```
tickTime=2000
initLimit=10
syncLimit=5
dataDir=/usr/local/zookeeper/zookeeper2/data
dataLogDir=/usr/local/zookeeper/zookeeper2/datalog
clientPort=2183
server.1=127.0.0.1:2888:3888
server.2=127.0.0.1:2889:3889
server.3=127.0.0.1:2890:3890
maxClientCnxns=60
```

⑤ 进入对应的 bin 目录,分别执行 "./zkServer.sh start" 命令,启动节点的 ZooKeeper 服务,如果看到 "Starting zookeeper ... STARTED" 返回信息,则表示 ZooKeeper 服务启动成功,操作命令及结果如下。

```
[root@master1 bin]# ./zkServer.sh start
ZooKeeper JMX enabled by default
Using config: /usr/local/zookeeper/zookeeper/bin/../conf/zoo.cfg
Starting zookeeper ... STARTED
[root@master1 bin]# /usr/local/zookeeper/zookeeper1/bin/zkServer.sh start
ZooKeeper JMX enabled by default
Using config: /usr/local/zookeeper/zookeeper1/bin/../conf/zoo.cfg
Starting zookeeper ... STARTED
[root@master1 bin]# /usr/local/zookeeper/zookeeper2/bin/zkServer.sh start
```

```
ZooKeeper JMX enabled by default
Using config: /usr/local/zookeeper/zookeeper2/bin/../conf/zoo.cfg
Starting zookeeper ... STARTED
```

⑥ 分别进入对应的 bin 目录，执行 "./zkServer.sh status" 命令，查询当前伪集群的状态，了解当前运行模式，操作命令及结果如下。

```
[root@master1 bin]# ./zkServer.sh status
ZooKeeper JMX enabled by default
Using config: /usr/local/zookeeper/zookeeper/bin/../conf/zoo.cfg
Mode: follower
[root@master1 bin]# ./zkServer.sh status
ZooKeeper JMX enabled by default
Using config: /usr/local/zookeeper/zookeeper1/bin/../conf/zoo.cfg
Mode: leader
[root@master1 bin]# ./zkServer.sh status
ZooKeeper JMX enabled by default
Using config: /usr/local/zookeeper/zookeeper2/bin/../conf/zoo.cfg
Mode: follower
```

3. ZooKeeper 集群模式搭建

在进行 ZooKeeper 集群模式搭建前，最好将伪集群模式的配置删除。

（1）环境准备

① 查看 Docker 容器，操作命令及结果如下。

微课 3-2
Zookeeper 的安装与配置 2

```
#查看 Docker 容器
[root@CentOS ~]# docker ps -a
CONTAINER ID  IMAGE    COMMAND  CREATED      STATUS        PORTS  NAMES
c3811fde5c00  slave2:2.0  "init"  4 weeks ago  Exited (137) 4
seconds ago   slave2
1a3ed0b818d6  slave1:2.0  "init"  4 weeks ago  Exited (137) 4
seconds ago   slave1
26fbe6d70f78  master1:1.0 "init"  4 weeks ago  Exited (137) 4
seconds ago                       master1
```

② 停止 Docker 服务，并查看容器的详细信息，操作命令及结果如下。

```
[root@CentOS ~]# systemctl stop docker
Warning: Stopping docker.service, but it can still be activated by:
 docker.socket


[root@CentOS ~]# cd /var/lib/docker/containers/
[root@CentOS containers]# ll
total 0
drwx--x---. 4 root root 237 Apr  5 11:38
1a3ed0b818d63fd1902bae9fd0b292d1e5178332844860e6cde1b18fea314053
drwx--x---. 4 root root 237 Apr  5 11:38
```

```
26fbe6d70f785e5c04dda8a7cc44c976999e30546f72f138209601ad1b9bf422
drwx--x---. 4 root root 237 Apr  5 11:38
c3811fde5c00ce3dc1051506674020af5284a0910b47ed974a1d32b29c9a918f
```

③ 为master1节点（ID以26fbe6d70f78开头）添加端口映射，操作命令及结果如下。

```
[root@CentOS containers]# cd
26fbe6d70f785e5c04dda8a7cc44c976999e30546f72f138209601ad1b9bf422/
[root@CentOS
26fbe6d70f785e5c04dda8a7cc44c976999e30546f72f138209601ad1b9bf422]# ls
26fbe6d70f785e5c04dda8a7cc44c976999e30546f72f138209601ad1b9bf422-json.log
checkpoints  config.v2.json  hostconfig.json  hostname  hosts  mounts
resolv.conf  resolv.conf.hash
```

编辑config.v2.json文件，找到"ExposedPorts"配置项，配置格式如下。

```
"ExposedPorts":{"容器的端口号/tcp":{}}
```

操作命令及配置参数值如下。

```
[root@CentOS 26fbe6d70f785e5c04dda8a7cc44c976999e30546f72f138209601ad1b9bf422]
# vi config.v2.json

"ExposedPorts":{"2181/tcp":{},"50070/tcp":{},"8088/tcp":{}}
```

配置结束后，保存并退出。

编辑hostconfig.json文件，找到"PortBindings"配置项，配置格式如下。

```
"PortBindings":{"容器的端口号/tcp":[{"HostIp":"","HostPort":"映射的宿主机端口"}]}
```

操作命令及配置参数值如下。

```
[root@CentOS 26fbe6d70f785e5c04dda8a7cc44c976999e30546f72f138209601ad1b9bf422]
# vi hostconfig.json

"PortBindings":{"2181/tcp":[{"HostIp":"","HostPort":"2181"}],"50070/tcp":
[{"HostIp":"","HostPort":"50070"}],"8088/tcp":[{"HostIp":"","HostPort":"8088"
}]}
```

配置结束后，保存并退出。

④ 参照为master1节点添加端口映射的方法，在slave1和slave2节点上依次编辑config.v2.json和hostconfig.json文件，为slave1和slave2节点添加端口映射。操作结束后，使用"docker ps -a"命令唤醒Docker服务，查看对应的端口映射，操作命令及结果如下。

```
[root@CentOS ~]# docker ps -a
CONTAINER ID   IMAGE       COMMAND      CREATED       STATUS
PORTS                                   NAMES

c3811fde5c00   slave2:2.0   "init"      4 weeks ago   Up 6 seconds
0.0.0.0:2183->2181/tcp, :::2183->2181/tcp, 0.0.0.0:8090->8088/tcp, :::8090->
8088/tcp                                slave2

1a3ed0b818d6   slave1:2.0   "init"      4 weeks ago   Up 7 seconds
```

项目 3　ZooKeeper 环境搭建与应用

```
0.0.0.0:2182->2181/tcp, :::2182->2181/tcp, 0.0.0.0:8089->8088/tcp,
:::8089->8088/tcp                                          slave1

26fbe6d70f78      master1:1.0    "init"          4 weeks ago   Up 7 seconds
0.0.0.0:2181->2181/tcp, :::2181->2181/tcp, 0.0.0.0:8088->8088/tcp,
:::8088->8088/tcp,0.0.0.0:50070->50070/tcp, :::50070->50070/tcp,   master1
```

（2）解压和配置文件

ZooKeeper 运行需要配置 JDK。（项目 1 中已经将 JDK 配置好，此处可以检查 JDK 的安装配置情况，确保没有问题，具体操作省略。）

① 在 master1 节点上，将准备好的 ZooKeeper 安装包解压到指定目录中，操作命令如下。

```
[root@master1 opt]# tar -zxvf zookeeper-3.4.8.tar.gz -C /usr/local/
```

② 进入 master1 节点的解压目录，修改目录名称，操作命令及结果如下。

```
[root@master1 opt]# cd /usr/local/
[root@master1 local]# ls
bin  games  lib64  share  src  etc  hadoop  include  lib  libexec  sbin
zookeeper-3.4.8
[root@master1 local]# mv zookeeper-3.4.8 zookeeper
```

③ 执行"vi ~/.bashrc"命令，修改 master1 节点的~/.bashrc 配置文件，修改完成后，按"Esc"键返回命令行模式，通过输入":wq!"保存修改并退出，随后执行"source ~/.bashrc"命令，使配置文件立即生效，操作命令及结果如下。

```
[root@master1 bin]# vi ~/.bashrc
export ZOOKEEPER_HOME=/usr/local/zookeeper
PATH=$PATH:$ZOOKEEPER_HOME/bin
[root@master1 bin]# source ~/.bashrc
```

④ 在 master1 节点的 ZooKeeper 目录中创建 data 和 datalog 目录，进入 data 目录并创建 myid 文件，文件内容为数字 1，操作命令及结果如下。

```
[root@master1 zookeeper]# mkdir data
[root@master1 zookeeper]# mkdir datalog
[root@master1 zookeeper]# cd data
[root@master1 data]# vi myid
[root@master1 data]# ls
myid
```

（3）配置 zoo.cfg 文件

① 重命名 3 个节点中的 zoo_sample.cfg 配置文件，操作命令（此处只展示对 master1 节点的操作命令）如下。

```
[root@master1 zookeeper]# cd conf
[root@master1 conf]# ls
configuration.xsl  log4j.properties  zoo_sample.cfg
[root@master1 conf]# mv zoo_sample.cfg zoo.cfg
```

② 打开并编辑 3 个节点中的 zoo.cfg 文件，这 3 个节点的 zoo.cfg 文件内容均相同，文

件内容如下。
```
tickTime=2000
initLimit=10
syncLimit=5
dataDir=/usr/local/zookeeper/data
dataLogDir=/usr/local/zookeeper/datalog
clientPort=2181
server.1=172.18.0.2:2887:3887
server.2=172.18.0.3:2888:3888
server.3=172.18.0.4:2889:3889
```
（4）集群分发（免密登录在项目1中已经配置）

在master1节点上，分发ZooKeeper安装目录和环境变量，操作命令及结果如下。
```
[root@master1 ~]# scp -r /usr/local/zookeeper slave1:/usr/local/
[root@master1 ~]# scp -r /usr/local/zookeeper slave2:/usr/local/
[root@master1 ~]# scp -r ~/.bashrc slave2:~/.bashrc
.bashrc                         100%  939     2.2MB/s   00:00
[root@master1 ~]# scp -r ~/.bashrc slave1:~/.bashrc
.bashrc                         100%  939     1.9MB/s   00:00
```
进入slave1节点，将myid内容修改为2，使配置文件生效。
```
[root@slave1 ~]# cd /usr/local/zookeeper/data
[root@slave1 data]# vi myid
[root@slave1 ~]# source ~/.bashrc
```
进入slave2节点，将myid内容修改为3，使配置文件生效。
```
[root@slave2 ~]# cd /usr/local/zookeeper/data
[root@slave2 data]# vi myid
[root@slave2 ~]# source ~/.bashrc
```
（5）测试安装

① 分别在3个节点的zookeeper/bin目录中执行"./zkServer.sh start"命令，启动集群，操作命令及结果如下。
```
[root@master1 bin]# ./zkServer.sh start
ZooKeeper JMX enabled by default
Using config: /usr/local/zookeeper/bin/../conf/zoo.cfg
Starting zookeeper ... STARTED

[root@slave1 bin]# ./zkServer.sh start
ZooKeeper JMX enabled by default
Using config: /usr/local/zookeeper/bin/../conf/zoo.cfg
Starting zookeeper ... STARTED

[root@slave2 bin]# ./zkServer.sh status
ZooKeeper JMX enabled by default
```

项目 3　ZooKeeper 环境搭建与应用

```
Using config: /usr/local/zookeeper/bin/../conf/zoo.cfg
Mode: follower
```

② 在任意节点的 zookeeper/bin 目录中执行 "./zkCli.sh -server 容器的 IP 地址" 命令，测试 ZooKeeper 节点与服务器端是否连接成功。如果在返回信息中看到 "Welcome to ZooKeeper!"，则表示 ZooKeeper 节点与服务器端连接成功，操作命令及部分结果如下。

```
[root@master1 bin]# ./zkCli.sh -server 172.18.0.2
.......
2023-04-06 11:04:12,826 [myid:] - INFO  [main:Environment@100] - Client environment:user.dir=/usr/local/zookeeper/bin
2023-04-06 11:04:12,826 [myid:] - INFO  [main:ZooKeeper@438] - Initiating client connection, connectString=172.18.0.2 sessionTimeout=30000 watcher=org.apache.zookeeper.ZooKeeperMain$MyWatcher@68de145
Welcome to ZooKeeper!
JLine support is enabled
#省略部分信息
2023-04-06 11:04:12,901 [myid:] - INFO  [main-SendThread(172.18.0.2:2181):ClientCnxn$SendThread@1299] - Session establishment complete on server 172.18.0.2/172.18.0.2:2181, sessionid = 0x1875483bbc40000, negotiated timeout = 30000

WATCHER::

WatchedEvent state:SyncConnected type:None path:null
[zk: 172.18.0.2(CONNECTED) 0]
```

③ 连接服务器并执行所有操作后，可以执行 "./zkServer.sh stop" 命令，停止 ZooKeeper 服务。

```
[root@master1 bin]# ./zkServer.sh stop
ZooKeeper JMX enabled by default
Using config: /usr/local/zookeeper/bin/../conf/zoo.cfg
Stopping zookeeper ... STOPPED
```

4. ZooKeeper 中监视机制的应用

（1）启动 ZooKeeper 集群

① 分别进入 3 个节点容器的 zookeeper/bin 目录，执行 "./zkServer.sh start" 命令，启动 ZooKeeper 集群，操作命令及结果（此处只展示 master1 节点的结果）如下。

```
[root@master1 bin]# ./zkServer.sh start
ZooKeeper JMX enabled by default
Using config: /usr/local/zookeeper/bin/../conf/zoo.cfg
Starting zookeeper ... STARTED
```

② 分别在 3 个节点上执行 "./zkServer.sh status" 命令，查看 ZooKeeper 节点的状态，操作命令及结果如下。

```
[root@master1 bin]# ./zkServer.sh status
ZooKeeper JMX enabled by default
Using config: /usr/local/zookeeper/bin/../conf/zoo.cfg
Mode: follower

[root@slave1 bin]# ./zkServer.sh status
ZooKeeper JMX enabled by default
Using config: /usr/local/zookeeper/bin/../conf/zoo.cfg
Mode: leader

[root@slave2 bin]# ./zkServer.sh status
ZooKeeper JMX enabled by default
Using config: /usr/local/zookeeper/bin/../conf/zoo.cfg
Mode: follower
```

（2）ZooKeeper 观察者模式

① 在本地进入 zookeeper/bin 目录，执行 "./zkCli.sh" 命令，连接本地服务器端，操作命令及结果如下。

```
[root@master1 bin]# ./zkCli.sh
Welcome to ZooKeeper!
JLine support is enabled
2023-04-06 11:11:50,861 [myid:] - INFO  [main-SendThread(localhost:2181):ClientCnxn$SendThread@1032] - Opening socket connection to server localhost/127.0.0.1:2181. Will not attempt to authenticate using SASL (unknown error)
2023-04-06 11:11:50,864 [myid:] - INFO  [main-SendThread(localhost:2181):ClientCnxn$SendThread@876] - Socket connection established to localhost/127.0.0.1:2181, initiating session
2023-04-06 11:11:50,870 [myid:] - INFO  [main-SendThread(localhost:2181):ClientCnxn$SendThread@1299] - Session establishment complete on server localhost/127.0.0.1:2181, sessionid = 0x1875483bbc40001, negotiated timeout = 30000

WATCHER::

WatchedEvent state:SyncConnected type:None path:null
[zk: localhost:2181(CONNECTED) 0]
```

② 在本地客户端中，执行 "ls / watch" 命令，查看根目录节点，发现只有一个节点处于观察者模式，操作命令及结果如下。

```
[zk: localhost:2181(CONNECTED) 0] ls / watch
[zookeeper]
```

③ 在其他两个节点中的任意一个节点的客户端上执行 "create /node 123" 命令，创建一

项目 3　ZooKeeper 环境搭建与应用

个新节点，操作命令及结果如下。

```
[zk: localhost:2181(CONNECTED) 0] create /node 123
Created /node
```

此时，在本地客户端的观察者模式下，可以监视到根目录中节点发生的变化，其监视日志如下。

```
[zk: localhost:2181(CONNECTED) 2]
WATCHER::

WatchedEvent state:SyncConnected type:NodeChildrenChanged path:/
```

④ 在创建 node 123 节点的客户端上，执行 "create /node2 123" 命令，再次创建一个新节点，操作命令及结果如下。

```
[zk: localhost:2181(CONNECTED) 1] create /node2 123
Created /node2
```

此时，在本地客户端的观察者模式下，查看监视日志，发现监视不到根目录中节点的变化，CONNECTED 后的数字仍是 2，没有新的监视日志提示，其监视日志如下。

```
[zk: localhost:2181(CONNECTED) 2]
WATCHER::

WatchedEvent state:SyncConnected type:NodeChildrenChanged path:/
```

注意　观察者模式具有这样的特性：监视命令被使用一次，只会监视一次，监视到了就输出内容，输出结束就退出监视。所以，根目录中节点的再次变化不会被监视到。

⑤ 连接本地服务器端之后，执行 "-help" 命令，查看帮助信息，操作命令及结果如下。

```
[zk: localhost:2181(CONNECTED) 3] -help
ZooKeeper -server host:port cmd args
    stat path [watch]
    set path data [version]
    ls path [watch]
    delquota [-n|-b] path
    ls2 path [watch]
    setAcl path acl
    setquota -n|-b val path
    history
    redo cmdno
    printwatches on|off
    delete path [version]
    sync path
    listquota path
    rmr path
    get path [watch]
    create [-s] [-e] path data acl
```

```
addauth scheme auth
quit
getAcl path
close
connect host:port
```

监视是 ZooKeeper 用来实现分布式锁、分布式配置管理、分布式队列等应用的主要手段。要想监视 ZooKeeper 文件系统中任意节点的变化（节点本身的增加、删除、数据的修改，以及子节点的变化），都可以通过在获取该数据时注册一个监视来实现。

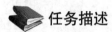 ZooKeeper CLI 操作

📚 任务描述

学习 ZooKeeper CLI 操作（包括创建节点、读取节点、设置数据、移除和删除节点等操作）及其相关知识。

📚 任务目标

（1）学会创建节点。
（2）学会从节点中读取数据。
（3）学会为节点设置数据。
（4）学会节点的移除和删除操作。

📚 知识准备

ZooKeeper CLI 用于与 ZooKeeper 集群进行交互。要执行 ZooKeeper CLI 操作，需要先进入节点的 zookeeper/bin 目录，执行"./zkServer.sh start"命令，启动 ZooKeeper 集群服务；再执行"./zkCli.sh"命令，使客户端连接 ZooKeeper 服务器。

1. 创建节点

创建节点的语法格式为 create [-s] [-e] path data acl。其中，"path"用于指定路径，由于 ZooKeeper 是树形结构的，所以创建"path"就是创建"path"节点；"data"是节点对应的数据，节点可以保存少量的数据；"acl"用来进行权限控制；"-s"用于指定创建的节点类型为顺序节点；"-e"用于指定创建的节点类型为临时节点。默认情况下，所有节点都是持久节点，当会话过期或客户端断开连接时，临时节点将被自动删除。其应用实例如下。

```
//在根目录中创建持久节点"FirstZnode"，用于存储数据"Myfirstzookeeper-app"
[zk:localhost:2181(CONNECTED)0]create /FirstZnode "Myfirstzookeeper-app"
//执行"ls /"命令查看根目录中的节点，并查看节点创建情况
[zk:localhost:2181(CONNECTED)1]ls /
```

指定参数为-s，创建顺序节点。

```
//在根目录中创建顺序节点"FirstZnode"，用于存储数据"second-data"
[zk:localhost:2181(CONNECTED)0]create -s /FirstZnode "second-data"
//执行"ls /"命令查看根目录中的节点，并查看节点创建情况
[zk:localhost:2181(CONNECTED)1]ls /
```

指定参数为-e，创建临时节点。

```
//在根目录中创建临时节点"SecondZnode"，用于存储数据"Ephemeral-data"
[zk:localhost:2181(CONNECTED)0]create -e /SecondZnode "Ephemeral-data"
//执行"ls /"命令查看根目录中的节点，并查看节点创建情况
[zk:localhost:2181(CONNECTED)1]ls /
```

2．读取节点

与读取相关的命令有"ls"和"get"。"ls"命令可以列出 ZooKeeper 指定节点的所有子节点，但是只能查看指定节点的第一级的所有子节点；"get"命令可以读取 ZooKeeper 指定节点的数据内容和属性信息。相关命令的语法格式分别为 ls path [watch]、get path [watch]、ls2 path [watch]。其中，"ls2"与"ls"不同的是，它可以查看到 time、version 等信息。其应用实例如下。

```
//读取根节点的所有子节点
[zk:localhost:2181(CONNECTED)0]ls /
[zk:localhost:2181(CONNECTED)1]ls2 /
//读取"FirstZnode"的数据内容和属性信息
[zk:localhost:2181(CONNECTED)2]get /FirstZnode
```

访问顺序节点时，必须输入节点的完整路径（注意顺序节点的命名）。

```
//读取顺序节点"FirstZnode0000000023"的数据内容和属性信息
[zk:localhost:2181(CONNECTED)3]get /FirstZnode0000000023
```

3．设置数据

使用"set"命令可以设置指定节点的数据，其语法格式为 set path data [version]。其中，"data"是要更新的内容，"version"表示数据版本。其应用实例如下。

```
//将 FirstZnode 节点的数据更新为 Data-updated
[zk:localhost:2181(CONNECTED)4]set /FirstZnode Data-updated
```

完成设置数据操作后，可以使用"get"命令检查数据。如果在"get"命令中设置了 watch 选项，则输出信息中将包含以下类似内容："WatchedEventstate:SyncConnectedtype: NodeDataChanged"。

4．创建子节点

创建子节点类似于创建新的节点，其区别在于子节点的路径需包含父节点的路径，语法格式为 create /parent_path/subnode_path data。其应用实例如下。

```
//创建 FirstZnode 的子节点
[zk:localhost:2181(CONNECTED)0]create /FirstZnode/Child1 firstchild
//查看 FirstZnode 的子节点
[zk:localhost:2181(CONNECTED)1]ls /FirstZnode
```

5. 检查状态

状态描述的是节点的元数据,它包含时间戳、版本号、操作控制列表、数据长度和子节点等属性,其语法格式为 stat /path。其应用实例如下。

```
//检查根目录中 FirstZnode 的节点状态
[zk:localhost:2181(CONNECTED)2]stat /FirstZnode
```

6. 移除节点

移除操作可移除指定的节点及其所有子节点,该操作需要在节点可用的情况下进行,其语法格式为 rmr /path。其应用实例如下。

```
//移除根目录中的 FirstZnode 节点及其子节点
[zk:localhost:2181(CONNECTED)3]rmr /FirstZnode
```

7. 删除节点

删除节点的语法格式为 delete /path。此命令类似于"rmr"命令,但它只适用于删除没有子节点的节点。

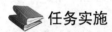

任务实施

1. 启动相关服务

① 进入 3 个节点的 zookeeper/bin 目录,执行"./zkServer.sh start"命令,启动服务,操作命令(此处只展示对 master1 节点的操作命令)及结果如下。

```
[root@master1 bin]# ./zkServer.sh start
ZooKeeper JMX enabled by default
Using config: /usr/local/zookeeper/bin/../conf/zoo.cfg
Starting zookeeper ... STARTED
```

② 在 master1 节点的 zookeeper/bin 目录下执行"jps"命令,查看系统进程,可以看到"QuorumPeerMain"进程,表示 ZooKeeper 已经启动,也可以执行"./zkServer.sh status"命令进行进程信息查看,操作命令及结果如下。

```
[root@master1 bin]# jps
217 QuorumPeerMain
330 Jps
[root@master1 bin]# ./zkServer.sh status
ZooKeeper JMX enabled by default
Using config: /usr/local/zookeeper/bin/../conf/zoo.cfg
Mode: follower
```

③ 打开客户端窗口,在服务端开启的情况下,执行"./zkCli.sh"命令,启动客户端并连接服务器,操作命令及部分结果如下。

```
[root@master1 bin]# ./zkCli.sh
Welcome to ZooKeeper!
JLine support is enabled
#省略部分信息
```

```
2023-04-06 11:11:50,870 [myid:] - INFO [main-SendThread(localhost:2181):
ClientCnxn$SendThread@1299] - Session establishment complete on server
localhost/127.0.0.1:2181, sessionid = 0x1875483bbc40001, negotiated timeout =
30000

WATCHER::

WatchedEvent state:SyncConnected type:None path:null
[zk: localhost:2181(CONNECTED) 0]
```

④ 若要连接不同的主机,则可执行"./zkCli.sh -server 容器主机名"命令,操作命令及结果如下。

```
[root@master1 bin]# ./zkCli.sh -server slave2
Connecting to slave2
#省略部分信息
2023-04-06 11:17:51,184 [myid:] - INFO [main:ZooKeeper@438] - Initiating client
connection, connectString=slave2 sessionTimeout=30000 watcher=org.apache.
zookeeper.ZooKeeperMain$MyWatcher@68de145
Welcome to ZooKeeper!
2023-04-06 11:17:51,218 [myid:] - INFO [main-SendThread(slave2.hadoop:2181):
ClientCnxn$SendThread@1299] - Session establishment complete on server
slave2.hadoop/172.18.0.4:2181, sessionid = 0x3875488a2830000, negotiated timeout
= 30000

WATCHER::

WatchedEvent state:SyncConnected type:None path:null
[zk: slave2(CONNECTED) 0]
```

⑤ 使用帮助命令"-help"来查看支持的操作,操作命令及结果如下。

```
[zk: localhost:2181(CONNECTED) 3] -help
ZooKeeper -server host:port cmd args
    stat path [watch]
    set path data [version]
    ls path [watch]
    delquota [-n|-b] path
    ls2 path [watch]
    setAcl path acl
    setquota -n|-b val path
    history
    redo cmdno
    printwatches on|off
    delete path [version]
    sync path
```

```
listquota path
rmr path
get path [watch]
create [-s] [-e] path data acl
addauth scheme auth
quit
getAcl path
close
connect host:port
```

2. 创建节点

（1）创建顺序节点

执行"create -s /zk-test 1001"命令，创建 zk-test 顺序节点，操作命令及结果如下。

```
[zk: localhost:2181(CONNECTED) 0] create -s /zk-test 1001
Created /zk-test0000000004
[zk: localhost:2181(CONNECTED) 1] ls /
[zookeeper, zk-test0000000004]
```

可以看到创建的 zk-test 顺序节点后面添加了 10 位数字序号以示区别。

（2）创建临时节点

① 执行"create -e /zk-temp 1001"命令，创建 zk-temp 临时节点，操作命令及结果如下。

```
[zk: localhost:2181(CONNECTED) 2] create -e /zk-temp 1001
Created /zk-temp
[zk: localhost:2181(CONNECTED) 3] ls /
[zookeeper, zk-test0000000004, zk-temp]
```

② 临时节点会在客户端会话结束后被自动删除。执行"quit"命令，退出客户端，再次使客户端连接服务器，并执行"ls /"命令，查看根目录中的节点，操作命令及结果如下。

```
[zk: localhost:2181(CONNECTED) 4] quit
Quitting...
2023-02-19 00:49:23,944 [myid:] - INFO  [main:ZooKeeper@684] - Session: 0x18667116eb90001 closed
2023-02-19 00:49:23,945 [myid:] - INFO  [main-EventThread:ClientCnxn$EventThread@519] - EventThread shut down for session: 0x18667116eb90001
[zk: localhost:2181(CONNECTED) 0] ls /
[zookeeper, zk-test0000000004]
```

可以看到根目录中已经不存在 zk-temp 临时节点。

（3）创建持久节点

执行"create /zk-permanent 1001"命令，创建 zk-permanent 持久节点，操作命令及结果如下。

```
[zk: localhost:2181(CONNECTED) 1] create /zk-permanent 1001
Created /zk-permanent
[zk: localhost:2181(CONNECTED) 2] ls /
[zk-permanent, zookeeper, zk-test0000000004]
```

项目 3　ZooKeeper 环境搭建与应用

可以看到持久节点不同于顺序节点，其节点名称后面不会被添加 10 位数字序列号。

3. 读取节点

① 读取根节点的所有子节点，执行"ls /"命令，操作命令及结果如下。

```
[zk: localhost:2181(CONNECTED) 2] ls /
[zk-permanent, zookeeper, zk-test0000000004]
```

② 读取根节点的数据内容和属性信息，执行"get /"命令，操作命令及结果如下。

```
[zk: localhost:2181(CONNECTED) 3] get /
cZxid = 0x0
ctime = Thu Jan 01 00:00:00 UTC 1970
mZxid = 0x0
mtime = Thu Jan 01 00:00:00 UTC 1970
pZxid = 0x40000000c
cversion = 11
dataVersion = 0
aclVersion = 0
ephemeralOwner = 0x0
dataLength = 0
numChildren = 3
```

也可以通过执行"ls2 /"命令来读取这些信息，操作命令及结果如下。

```
[zk: localhost:2181(CONNECTED) 4] ls2 /
[zk-permanent, zookeeper, zk-test0000000004]
cZxid = 0x0
ctime = Thu Jan 01 00:00:00 UTC 1970
mZxid = 0x0
mtime = Thu Jan 01 00:00:00 UTC 1970
pZxid = 0x40000000c
cversion = 11
dataVersion = 0
aclVersion = 0
ephemeralOwner = 0x0
dataLength = 0
numChildren = 3
```

③ 执行"get /zk-permanent"命令，读取 zk-permanent 节点的数据内容和属性信息，操作命令及结果如下。

```
[zk: localhost:2181(CONNECTED) 5] get /zk-permanent
1001
cZxid = 0x40000000c
ctime = Sun Feb 19 00:51:02 UTC 2023
mZxid = 0x40000000c
mtime = Sun Feb 19 00:51:02 UTC 2023
```

```
pZxid = 0x40000000c
cversion = 0
dataVersion = 0
aclVersion = 0
ephemeralOwner = 0x0
dataLength = 4
numChildren = 0
```

4. 更新节点

执行"set /zk-permanent 456"命令，将 zk-permanent 节点的数据内容更新为 456，可以看到此时 dataVersion 更新为 1，即表示数据进行了更新，操作命令及结果如下。

```
[zk: localhost:2181(CONNECTED) 6] set /zk-permanent 456
cZxid = 0x40000000c
ctime = Sun Feb 19 00:51:02 UTC 2023
mZxid = 0x40000000d
mtime = Sun Feb 19 00:55:20 UTC 2023
pZxid = 0x40000000c
cversion = 0
dataVersion = 1
aclVersion = 0
ephemeralOwner = 0x0
dataLength = 3
numChildren = 0
```

5. 删除节点

使用"delete"命令可以删除 ZooKeeper 的指定节点。

执行"delete /zk-permanent"命令，删除 zk-permanent 节点，使用"ls /"命令查看节点信息，操作命令及结果如下。

```
[zk: localhost:2181(CONNECTED) 7] delete /zk-permanent
[zk: localhost:2181(CONNECTED) 8] ls /
[zookeeper, zk-test0000000004]
```

从上述结果中可以看到，已经成功删除了 zk-permanent 节点。

若要删除的节点存在子节点，那么无法直接使用"delete"命令删除该节点，必须先删除其子节点，再删除该节点。

项目小结

本项目介绍了 ZooKeeper 的架构、组件、常见服务、工作流、功能和优势等，并重点介绍了 ZooKeeper 的选举机制、监视机制、ZooKeeper CLI 操作及其相关知识，完成了 ZooKeeper

项目 3　ZooKeeper 环境搭建与应用

的安装与配置等操作。通过对本项目的学习，读者可对 ZooKeeper 的相关技术与应用建立清晰的认识，并为使用 ZooKeeper 进行分布式集群管理奠定良好的基础。

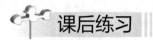

课后练习

1. ZooKeeper 集群中有哪些角色？
2. ZooKeeper 有几种部署模式？
3. 请简述 ZooKeeper 的工作原理。

项目 ④ HBase 环境搭建与基本操作

学习目标

【知识目标】
了解 HBase 的产生背景、HBase 架构。
识记 HBase 常用操作。

【技能目标】
学会 HBase 的安装与配置。
学会 HBase Shell 命令的使用。

【素质目标】
具有严谨细致的工作态度和工作作风。
具有良好的团队协作意识和业务沟通能力。
具有良好的表达能力和文档查阅能力。
具有规范的编程意识和较好的数据洞察力。

项目描述

HBase 是一个高可靠、高性能、面向列、可伸缩的分布式列存储数据库,是谷歌 BigTable 的开源实现,主要用来存储非结构化和半结构化的松散数据。HBase 的目标是处理庞大的表,其可以利用较为廉价的计算机集群,通过水平扩展的方式处理由数十亿行和数百万列的元素组成的数据表。

本项目主要引导学生完成 HBase 的安装与配置,学习 HBase Shell 的基本操作。

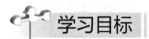   HBase 的安装与配置

任务描述

(1)学习 HBase 相关技术知识,了解 HBase 的产生背景、特点等。
(2)完成 HBase 的安装与配置。

任务目标

(1)熟悉 HBase 的特点。
(2)学会 HBase 的安装与配置。

项目 4　HBase 环境搭建与基本操作

知识准备

HBase 是建立在 HDFS 之上的分布式、面向列的数据库。它是一个开源项目，是可水平扩展的，可以快速随机访问海量结构化数据。它利用 HDFS 提供的容错能力，可以直接存储 HDFS 数据。可以使用 HBase 在 HDFS 中随机访问数据，如图 4-1 所示。HBase 建立在 HDFS 之上，并提供了读写访问功能。

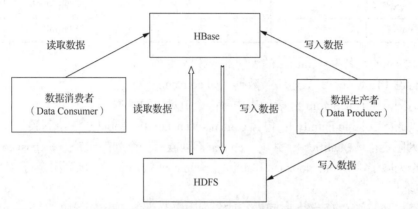

图 4-1　使用 HBase 在 HDFS 中随机访问数据

HBase 作为面向列的数据库运行在 HDFS 上，HDFS 缺乏随机读写操作，两者存在较大的区别。HDFS 和 HBase 的对照如表 4-1 所示。

表 4-1　HDFS 和 HBase 的对照

HDFS	HBase
HDFS 是适用于存储大容量文件的分布式文件系统	HBase 是建立在 HDFS 之上的数据库
HDFS 不支持对记录的快速单独查找	HBase 可在较大的表中进行快速查找
HDFS 提供了高延迟批量处理功能	HBase 提供了在数十亿条记录中低延迟访问单行记录（随机存取）的功能
HDFS 中的数据只允许被顺序访问	HBase 内部使用哈希表且提供随机接入功能，其可存储索引，对在 HDFS 文件中的数据进行快速查找

1. HBase 的存储机制

在 HBase 中，数据存储在具有行和列的表中。与关系数据库的区别是，关系数据库通过行和列确定要查找的值，而在 HBase 中通过行键、列（列族:列限定符）和时间戳来确定要查找的值。因此，在关系数据库的表中，值的映射关系是二维的，而在 HBase 表中，值的映射关系是多维的。下面通过例子来帮助读者理解 HBase 表的存储结构。其中，HBase 数据结构如表 4-2 所示。

表 4-2 HBase 数据结构

行键	时间戳	contents	anchor		people
		cnnsi.com	my.look.ca	html	
"com.cnn.www"	t9	"cnn"			
"com.cnn.www"	t8		"cnn.com"		
"com.cnn.www"	t6			"html…"	
"com.cnn.www"	t5			"html…"	
"com.cnn.www"	t3			"html…"	

表 4-2 描述的是名为 webtable 的表的部分数据。

（1）行键（RowKey）：表示一行数据，如 com.cnn.www。

（2）时间戳（TimeStamp）：表示表中每一个值对应的版本。

（3）列族（Column Family）：包含 contents、anchor 和 people 这 3 个列族。

（4）列限定符（Qualifier）：其中，contents 列族有一个列限定符——cnnsi.com；anchor 列族有两个列限定符——my.look.ca 和 html；people 列族是空列，无列限定符，即在 HBase 中没有数据。

（5）值（Value）：由{行键，时间戳，列族:列限定符}联合确定。例如，值 cnn 由{com.cnn.www，t9，contents:cnnsi.com}联合确定。

下面通过概念视图和物理视图来进一步讲述 HBase 表的存储结构。

2. 概念视图

在 HBase 中，从概念层面上讲，表 4-2 中展现的是由一组稀疏的行组成的表，期望按列族（contents、anchor 和 people）进行物理存储，并且可随时将新的列限定符（cnnsi.com、my.look.ca、html 等）添加到现有的列族中。每一个值都对应一个时间戳，每行行键中的值相同。可以将这样的表想象成一个大的映射关系，通过行键、行键+时间戳或行键+列（列族:列限定符）就可以定位指定的数据。可以把这种关系用概念视图来表示，如表 4-3 所示。

表 4-3 Hbase 表的概念视图

行键	时间戳	contents	anchor
"com.cnn.www"	t9	contents:cnnsi.com="cnn"	
	t8		anchor:my.look.ca="cnn.com"
	t6		anchor:html="<html>..."
	t5		anchor:html="<html>..."
	t3		anchor:html="<html>..."

在解释表 4-3 所示的概念视图前，先来熟悉以下术语。

（1）表（Table）：由一个或多个分区表组成。分区表按列进行分区，每个分区表都有自己的列族和行键。分区表使得 HBase 可以在不同的机器上存储数据，从而实现 HBase 的高伸缩性和高可用性。

项目 4　HBase 环境搭建与基本操作

（2）行键：行键是数据行在表中的唯一标志，并作为检索记录的主键。表 4-3 中，"com.cnn.www" 就是行键的值。一个表中会有若干个行键，且行键的值不能重复。行键按字典顺序排列，排列顺序是从最小值到最大值。按行键检索数据可以有效地减少查询时间。在 HBase 中，可以用以下 3 种方式访问表中的行。

① 通过单个行键访问。
② 通过给定行键的范围访问。
③ 进行全表扫描。

行键可以用任意字符串（字符串最大长度为 64KB）表示并按照字典顺序进行存储。对于经常一起读取的行，可以对行键的值进行精心设计，以便将它们放在一起存储。

（3）列族：列族和表一样需要在架构表时预先声明，列族的前缀必须由可输出的字符组成。从物理上讲，所有列族成员一起被存储在文件系统中。HBase 中的列限定符被分组到列族中，不需要在架构时定义，可以在表启动并运行时动态变换。例如，表 4-2 中的 contents 和 anchor 是列族，而它们对应的列限定符（cnnsi.com、my.look.ca、html）在插入值时定义即可。

（4）单元格（Cell）：一个 {Row，Column，Version} 元组精确地指定了 HBase 中的一个单元格。其中，行（Row）包含一个行键、多个列（Column）以及对应的值。列由列族和列限定符两部分组成，列族不能随意增减，列限定符可以动态增加。每个单元格可以存储多个版本（Version）的数据，可以使用时间戳作为版本号来区分不同版本。

（5）时间戳：默认取平台时间，也可自定义时间，它是单元中指定的多个版本值中的某个版本值的标志。例如，在表 4-3 中，由 {com.cnn.www,contents:html} 确定 3 个值，这 3 个值可以被称为值的 3 个版本，而这 3 个版本分别对应的时间戳的值为 t6、t5、t3。

（6）值（Value）：是单元中存储的具体数据内容，由 {Row，Column，Version} 确定。

3. 物理视图

观察表 4-2 和表 4-3 可以发现，表 4-2 中 people 列族在表 4-3 中并没有体现，这是因为在 HBase 中没有值的单元格并不占用内存空间。HBase 的存储形式是按照列存储的稀疏行/列矩阵，物理视图实际上是对概念视图中的行进行切割，并按照列族进行存储。

表 4-3 所示的概念视图对应的物理视图（即 HBase 内部物理存储的实现形式）如表 4-4 所示。

表 4-4　Hbase 表的物理视图

行键	时间戳	列族
"com.cnn.www"	t9	contents:cnnsi.com="cnn"
	t8	anchor:my.look.ca="cnn.com"
	t6	anchor:html="<html>..."
	t5	anchor:html="<html>..."
	t3	anchor:html="<html>..."

从表 4-4 中可以看出，空值是不被存储的，所以查询时间戳为 t8 的 "contents: html" 将返回 null，同样，查询时间戳为 t9 的 "anchor:my.look.ca" 也会返回 null。如果没有指明

时间戳，那么应该返回指定列的最新数据值，并且最新数据值在表中是最先找到的，因为它们是按照时间降序排列的。所以，如果查询"contents:"而不指明时间戳，那么将返回时间戳为 t6 的数据；如果查询"anchor"的"my.look.ca"而不指明时间戳，那么将返回时间戳为 t8 的数据。这种存储结构还可以随时向 HBase 表中的任何一个列族添加新列，而不需要事先说明。

HBase 的表中的一行由若干列组成，若干列又构成一个列族，这不仅有助于构建数据的语义边界或者局部边界，还有助于给它们设置某些特性（如压缩），或者指示它们如何存储在内存中。一个列族的所有列存储在同一个底层的存储文件中，这个存储文件称为 HFile。所有的行按照行键字典顺序进行排序存储。

列式数据库与行式数据库（传统数据库的一种形式）的比较如表 4-5 所示。

表 4-5 列式数据库与行式数据库的比较

列式数据库	行式数据库
其适用于在线分析处理	其适用于联机事务处理
列式数据库被设计为巨大的表	行式数据库被设计为小数目的行和列

HBase 和关系数据库管理系统（Relational Database Management System，RDBMS）的比较如表 4-6 所示。

表 4-6 HBase 和 RDBMS 的比较

HBase	RDBMS
HBase 无模式，不具有固定列模式的概念；仅定义列族	RDBMS 有模式，描述了表的整体结构的约束
其专门创建为宽表，且是可水平扩展的	其创建为细而小的表，很难形成规模
没有任何事务存于 HBase 中	RDBMS 是事务性的
其一般用于处理非规范化的数据	其一般用于处理具有规范化的数据
其对于半结构化及结构化数据的处理效果较好	其对于结构化的数据处理效果较好

4．HBase 架构

从物理上说，HBase 是由 3 种类型的服务器以主从模式构成的。这 3 种类型的服务器分别是 HRegion Server、HMaster 和 ZooKeeper。其中，HRegion Server 负责数据的读写，用户通过 HRegion Server 来实现对数据的访问；HMaster 负责对 HRegion Server 的管理，以及对数据库的创建和删除等操作；ZooKeeper 作为 HDFS 的一部分，负责维护集群中服务器的状态（某个服务器是否在线、服务器之间数据的同步操作及 Leader 的选举等）。另外，Hadoop DataNode 负责存储所有 HRegion Server 所管理的数据。HBase 中的所有数据都是以 HDFS 文件的形式存储的。出于使 HRegion Server 所管理的数据更加本地化的考虑，HRegion Server 是根据 DataNode 分布的。HBase 的数据在写入的时候都存储在本地。但当某一个 Region（HBase 中分布式存储和负载均衡的最小单元）被移除或被重新分配的时候，就可能产生数据不在本地的情况，在此对这种情况不做讨论。NameNode 负责维护构成文件的所有物理数据块的元信息。HBase 架构如图 4-2 所示。

项目 4　HBase 环境搭建与基本操作

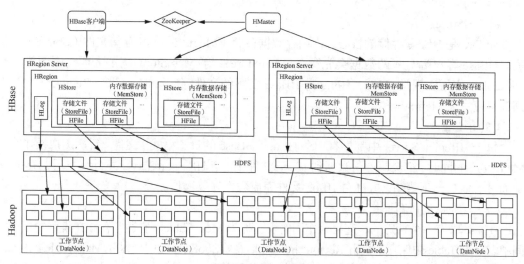

图 4-2　HBase 架构

（1）ZooKeeper

HBase 利用 ZooKeeper 维护集群中服务器的状态并协调分布式系统的工作。ZooKeeper 使用一致性算法来保证服务器之间的同步，并负责 Leader 选举的工作。需要注意的是，要想保证良好的一致性及顺利的 Leader 选举，集群中服务器的数目就必须是奇数。

（2）HMaster

HMaster 负责 Regions 的分配、数据库的创建和删除操作。具体来说，HMaster 的职责如下。

① 调控 HRegion Server 的工作。

② 在集群启动的时候分配 Regions，根据恢复服务或者负载均衡的需要重新分配 Regions。

③ 监控集群中 HRegion Server 的工作状态。

④ 管理数据库。

⑤ 提供创建、删除或者更新表的接口。

（3）HRegion Server

所有的数据库数据一般都被保存在 HDFS 中，用户通过一系列 HRegion Server 获取这些数据。一台机器上一般只运行一个 HRegion Server，且每一个区段的 HRegion 只会被一个 HRegion Server 维护。

HRegion Server 主要负责响应用户的 I/O 请求，并向 HDFS 读写数据，是 HBase 中最核心的模块之一。

HRegion Server 内部管理了一系列 HRegion 对象，每个 HRegion 对象都对应表中的一个 Region，一个 HRegion 由多个 HStore 组成。每个 HStore 都对应表中的一个列族的存储，每个列族其实就是一个集中的存储单元。

（4）HRegion

HBase 中的表是根据行键的值水平分割为 Region 的。一个 Region 包含表中所有行键位于 Region 的起始键值和结束键值之间的行。从物理上讲，最初 HBase 中会建立一个 Region，随着表记录数的增加，表内容所占资源将不断增加，当增加到指定阈值时，一个表将被拆分为两块，每一块就是一个 HRegion。以此类推，表随着记录数的不断增加而变大后，会逐渐拆分为若干个 HRegion。一张完整的表被保存在多个 HRegion 中。

（5）HStore

HStore 是 HBase 存储的核心，由内存数据存储（MemStore）和存储文件（StoreFile）组成。HBase 客户端写入的数据会先写入 MemStore，当 MemStore 满了之后会刷新为一个 StoreFile（底层实现是 HFile，HFile 是 HBase 实际的文件存储格式）。每个 Region 中的一个列族对应一个 MemStore。

（6）HLog

每个 HRegion Server 中都会有一个 HLog（HBase 的日志格式）文件，每次用户操作写入 MemStore 的同时，都会写一份数据到 HLog 文件中。当 HRegion Server 意外终止后，HMaster 会通过 ZooKeeper 感知，并通过 HLog 文件完成数据恢复操作。

（7）HBase 的读写操作

HBase 中有一个特殊的、起目录作用的表，称为 Meta Table，它通常存储在集群的根表空间中。Meta Table 中保存了集群 Regions 的地址信息。ZooKeeper 中会保存 Meta Table 的位置。

第一次对 HBase 进行读写操作将按照以下步骤执行。

① HBase 客户端从 ZooKeeper 中得到保存 Meta Table 的 HRegion Server 的信息。

② HBase 客户端向该 HRegion Server 查询负责管理需要访问的行键所在 Region 的 HRegion Server 地址。客户端会缓存这一信息及 Meta Table 所在位置的信息。

③ HBase 客户端与查询到的 HRegion Server 进行通信，实现对行的读写操作。在此后的读写操作中，HBase 客户端会根据缓存寻找相应的 HRegion Server 地址。当该 HRegion Server 不再可达时，HBase 客户端将会重新访问 Meta Table 并更新缓存。

5. HBase 的特点

（1）面向列

HBase 是面向列的存储和权限控制的数据库，列式存储的数据在表中是按照某列存储的，这种存储方式能够减少查询时的数据量，尤其在分析特定列时表现更佳。

（2）数据多版本

每个单元中的数据可以有多个版本，默认情况下，版本号自动分配，即单元格插入时的时间戳。

（3）具有稀疏性

值为空的列并不占用存储空间，因此表可以设计得非常稀疏。

（4）具有扩展性

HBase 底层依赖 HDFS，可借助 HDFS 的分布式存储实现水平扩展。

（5）具有高可靠性

HBase 的预写式日志（Write-Ahead Logging）机制保证了数据在写入时不会因集群异常而导致写入数据丢失；复制（Replication）机制保证了在集群出现严重的问题时，数据不会丢失或损坏。此外，由于 HBase 底层使用了 HDFS，而 HDFS 本身也有备份，因此，在一定程度上也提高了 HBase 的可靠性。

（6）具有高性能

底层的日志结构合并树（Log-Structured Merge Tree，LSM）数据结构和行键有序排列等架构上的独特设计，使 HBase 具有非常高的写入性能；Regions 拆分、主键索引和缓存机制使得 HBase 在海量数据下具备一定的随机读取性能，该性能使 HBase 针对行键的查询速度能达到毫秒级别。

项目 4　HBase 环境搭建与基本操作

📖 任务实施

HBase 支持单机模式部署和伪分布式模式部署，下面分别按照这两种模式对 HBase 进行部署。

1. HBase 单机模式部署

（1）HBase 的安装

① 准备好 HBase 的安装包后，进入存储目录，执行"tar -zxvf hbase-1.6.0-bin.tar.gz -C /usr/local/"命令，解压安装包，操作命令如下。

微课 4-1　HBase 的安装与配置 1

```
[root@master1 opt]$ tar -zxvf hbase-1.6.0-bin.tar.gz -C /usr/local/
```

② 进入解压目录，执行"mv hbase-1.6.0 ./hbase"命令，将解压后的安装包重命名为 hbase（以方便后续操作），操作命令如下。

```
[root@master local]$ mv hbase-1.6.0 ./hbase
```

③ 执行"chown -R hadoop:hadoop hbase"命令，修改 hbase 文件的权限，操作命令如下。

```
[root@master local]$ chown -R hadoop:hadoop hbase
```

（2）HBase 的配置

进入 hbase/conf 目录，编辑 hbase-site.xml 配置文件，文件内容如下。

```
<configuration>
<property>
<name>hbase.rootdir</name>
<value>file:///usr/local/hbase/logs/site</value>
</property>
<property>
<name>hbase.cluster.distributed</name>
<value>true</value>
</property>
</configuration>
```

（3）HBase 的启动

① 配置完成后，进入 hbase/bin 目录，执行"./start-hbase.sh"命令，启动 HBase，操作命令如下。

```
[root@master bin]$ ./start-hbase.sh
```

② 执行"jps"命令，查看系统进程，验证 HBase 单机模式部署成功，操作命令及结果如下。

```
[root@master bin]$ jps
528 HMaster
752 Jps
646 HRegionServer
```

2. HBase 伪分布式模式部署

（1）基础环境的准备

HBase 伪分布式模式下的数据设置存储在 HDFS 中，所以在进行 HBase 伪分布式模式的部署前，需要安装和配置 Hadoop 环境（详细安装过程请参考项目 1 的任务 1.3），通过执行"start-all.sh"命令启动 Hadoop 服务，并通过执

微课 4-2　HBase 的安装与配置 2

109

行"jps"命令查看Hadoop进程,操作命令及结果如下。

```
[root@master logs]$ jps
12418 NameNode
12901 ResourceManager
13189 NodeManager
13322 Jps
12555 DataNode
12732 SecondaryNameNode
```

(2)伪分布式环境搭建

① 进入安装包的存储目录,执行"tar -zxvf hbase-1.6.0-bin.tar.gz -C /usr/local"命令,将HBase解压到指定路径中,操作命令如下。

```
[root@master opt]$ tar -zxvf hbase-1.6.0-bin.tar.gz -C /usr/local
```

② 进入解压目录,执行"mv hbase-1.6.0 hbase"命令,进行重命名操作,执行"chown -R hadoop:hadoop hbase/"命令,修改hbase目录的所有者,操作命令如下。

```
[root@master local]$ mv hbase-1.6.0 hbase
[root@master local]$ chown -R hadoop:hadoop hbase/
```

③ 编辑~/.bashrc文件,将HBase的安装路径添加到环境变量中,操作内容如下。

```
export HBASE_HOME=/usr/local/hbase
PATH=$PATH:$HBASE_HOME/bin
```

④ 修改完成后,按"Esc"键返回命令行模式,通过输入":wq!"保存修改并退出,随后执行"source ~/.bashrc"命令,使配置文件立即生效,操作命令如下。

```
[root@master local]$ source ~/.bashrc
```

⑤ 执行"ll"命令,进入hbase/conf目录查看文件,可以看到conf目录中有配置文件hbase-env.sh和hbase-site.xml,操作命令及结果如下。

```
[root@master conf]$ ll
total 44
-rw-r--r--. 1 hadoop hadoop 1811 Oct 19  2019 hadoop-metrics2-hbase.properties
-rw-r--r--. 1 hadoop hadoop 4284 Oct 19  2019 hbase-env.cmd
-rw-r--r--. 1 hadoop hadoop 7603 Feb 25 08:54 hbase-env.sh
-rw-r--r--. 1 hadoop hadoop 2257 Oct 19  2019 hbase-policy.xml
-rw-r--r--. 1 hadoop hadoop  934 Feb 25 07:55 hbase-site.xml
-rw-r--r--. 1 hadoop hadoop 1168 Oct 19  2019 log4j-hbtop.properties
-rw-r--r--. 1 hadoop hadoop 4977 Oct 19  2019 log4j.properties
-rw-r--r--. 1 hadoop hadoop   10 Oct 19  2019 regionservers
```

⑥ 编辑配置文件hbase-env.sh,找到配置文件中对应的配置项,并修改配置项内容如下。

```
export JAVA_HOME=/usr/lib/jdk1.8.0
export HBASE_MANAGERS_ZK=true
```

⑦ 编辑配置文件hbase-site.xml,修改配置项内容如下。

```
<configuration>
<property>
<name>hbase.rootdir</name>
```

项目 4　HBase 环境搭建与基本操作

```
<value>hdfs://localhost:9000/hbase</value>
</property>
<property>
<name>hbase.cluster.distributed</name>
<value>true</value>
</property>
<property>
</configuration>
```

（3）启动 HBase

进入 /usr/local/hbase/bin 目录，执行 "./start-hbase.sh" 命令，启动 HBase，操作命令及结果如下。

```
[root@master bin]$ ./start-hbase.sh
SLF4J: Class path contains multiple SLF4J bindings.
SLF4J: Found binding in [jar:file:/usr/local/hadoop/share/hadoop/common/lib/
slf4j-log4j12-1.7.10.jar!/org/slf4j/impl/StaticLoggerBinder.class]
SLF4J: Found binding in [jar:file:/usr/local/hbase/lib/client-facing-
thirdparty/slf4j-log4j12-1.7.25.jar!/org/slf4j/impl/StaticLoggerBinder.class]
SLF4J: Actual binding is of type [org.slf4j.impl.Log4jLoggerFactory]
SLF4J: Class path contains multiple SLF4J bindings.
```

（4）查看系统进程

执行 "jps" 命令，查看系统进程，验证 HBase 伪分布式模式部署是否成功，操作命令及结果如下。

```
[root@master bin]$ jps
2832 HMaster
1078 HQuorumPeer
1270 HRegionServer
2151 ResourceManager
1800 DataNode
1660 NameNode
2332 NodeManager
1981 SecondaryNameNode
3039 Jps
```

3. 使用外置 ZooKeeper 搭建 HBase 集群

（1）基础环境的准备

在项目 1 中完成了 Hadoop 完全分布式环境的搭建，在项目 3 中完成了 ZooKeeper 环境的搭建。在此基础上继续进行如下操作，使用外置 ZooKeeper 搭建 HBase 集群。

① 在宿主机环境中执行 "docker ps -a" 命令，查看 Docker 容器，操作命令及结果如下。

```
[root@CentOS ~]# docker ps -a
CONTAINER ID     IMAGE          COMMAND         CREATED        STATUS
PORTS                           NAMES
```

```
c3811fde5c00      slave2:2.0    "init"          4 weeks ago       Exited (137) 4
seconds ago                     slave2
1a3ed0b818d6      slave1:2.0    "init"          4 weeks ago       Exited (137) 4
seconds ago                     slave1
26fbe6d70f78      master1:1.0   "init"          4 weeks ago       Exited (137) 4 seconds
ago                             master1
```

② 执行 "systemctl stop docker" 命令，停止 Docker 服务，执行 "cd /var/lib/docker/containers/" 进入 containers 目录，随后执行 "ll" 命令，查看容器详细信息，操作命令及结果如下。

```
[root@CentOS ~]# systemctl stop docker
Warning: Stopping docker.service, but it can still be activated by:
  docker.socket

[root@CentOS ~]# cd /var/lib/docker/containers/
[root@CentOS containers]# ll
total 0
drwx--x---. 4 root root 237 Apr  5 11:38 1a3ed0b818d63fd1902bae9fd0b292d1e5178332844860e6cde1b18fea314053
drwx--x---. 4 root root 237 Apr  5 11:38 26fbe6d70f785e5c04dda8a7cc44c976999e30546f72f138209601ad1b9bf422
drwx--x---. 4 root root 237 Apr  5 11:38 c3811fde5c00ce3dc1051506674020af5284a0910b47ed974a1d32b29c9a918f
```

③ 为 master1 节点（ID 以 26fbe6d70f78 开头）添加端口映射，操作命令如下。

微课 4-3　HBase 的安装与配置 3

```
[root@CentOS containers]# cd 26fbe6d70f785e5c04dda8a7cc44c976999e30546f72f138209601ad1b9bf422/
[root@CentOS 26fbe6d70f785e5c04dda8a7cc44c976999e30546f72f138209601ad1b9bf422]# ls
26fbe6d70f785e5c04dda8a7cc44c976999e30546f72f138209601ad1b9bf422-json.log
checkpoints  config.v2.json  hostconfig.json  hostname  hosts  mounts
resolv.conf  resolv.conf.hash
```

编辑 config.v2.json 文件，找到 "ExposedPorts" 配置项，配置格式如下。

```
"ExposedPorts":{"容器的端口号/tcp":{}}
```

操作命令及配置参数值如下。

```
[root@CentOS 26fbe6d70f785e5c04dda8a7cc44c976999e30546f72f138209601ad1b9bf422]# vi config.v2.json
```

```
"ExposedPorts":{"16010/tcp":{},"16020/tcp":{},"16030/tcp":{},"2181/tcp":{},"22/tcp":{},"50070/tcp":{},"8088/tcp":{}}
```

配置结束后，保存并退出，编辑 hostconfig.json 文件，找到 "PortBindings" 配置项，配置格式如下。

```
"PortBindings":{"容器的端口号/tcp":[{"HostIp":"","HostPort":"映射的宿主机端口"}]}
```

项目 4 HBase 环境搭建与基本操作

操作命令及配置参数值如下。

```
[root@CentOS 26fbe6d70f785e5c04dda8a7cc44c976999e30546f72f138209601ad1b9bf422]# vi hostconfig.json

"PortBindings":{"16010/tcp":[{"HostIp":"","HostPort":"16010"}],"16020/tcp":[{"HostIp":"","HostPort":"16020"}],"16030/tcp":[{"HostIp":"","HostPort":"16030"}],"2181/tcp":[{"HostIp":"","HostPort":"2181"}],"50070/tcp":[{"HostIp":"","HostPort":"50070"}],"8088/tcp":[{"HostIp":"","HostPort":"8088"}]}
```

配置结束后,保存并退出。

④ 按照为 master1 节点添加端口映射的方法,在 slave1 和 slave2 节点上依次编辑 config.v2.json 和 hostconfig.json 文件,为 slave1 和 slave2 节点添加端口映射。操作结束后,使用"docker ps -a"命令唤醒 Docker 服务,并查看对应的端口映射,操作命令及结果如下。

```
[root@CentOS ~]# docker ps -a
CONTAINER ID    IMAGE        COMMAND       CREATED       STATUS
PORTS                                                    NAMES

c3811fde5c00    slave2:2.0   "init"        4 weeks ago   Up 6 seconds
0.0.0.0:2183->2181/tcp, :::2183->2181/tcp, 0.0.0.0:8090->8088/tcp, :::8090->8088/tcp, 0.0.0.0:16012->16010/tcp, :::16012->16010/tcp, 0.0.0.0:16022->16020/tcp, :::16022->16020/tcp, 0.0.0.0:16032->16030/tcp, :::16032->16030/tcp
slave2

1a3ed0b818d6    slave1:2.0   "init"        4 weeks ago   Up 7 seconds
0.0.0.0:2182->2181/tcp, :::2182->2181/tcp, 0.0.0.0:8089->8088/tcp, :::8089->8088/tcp, 0.0.0.0:16011->16010/tcp, :::16011->16010/tcp, 0.0.0.0:16021->16020/tcp, :::16021->16020/tcp, 0.0.0.0:16031->16030/tcp, :::16031->16030/tcp
slave1

26fbe6d70f78    master1:1.0  "init"        4 weeks ago   Up 7 seconds
0.0.0.0:2181->2181/tcp, :::2181->2181/tcp, 0.0.0.0:8088->8088/tcp, :::8088->8088/tcp, 0.0.0.0:16010->16010/tcp, :::16010->16010/tcp, 0.0.0.0:16020->16020/tcp, :::16020->16020/tcp, 0.0.0.0:16030->16030/tcp, :::16030->16030/tcp, 0.0.0.0:50070->50070/tcp, :::50070->50070/tcp,                                 master1
```

上述操作主要实现的端口映射如下:Hadoop 端口 50070 和 8088,ZooKeeper 端口 2181,HBase 端口 16010、16020 和 16030。

在 master1 节点上执行"start-all.sh"命令,启动 Hadoop 服务,操作命令及结果如下。

```
[root@master1 /]# start-all.sh
This script is Deprecated. Instead use start-dfs.sh and start-yarn.sh
Starting namenodes on [master1]
master1: starting namenode, logging to /usr/local/hadoop/logs/hadoop-root-
```

```
namenode-master1.out
slave2: starting datanode, logging to /usr/local/hadoop/logs/hadoop-root-
datanode-slave2.out
slave1: starting datanode, logging to /usr/local/hadoop/logs/hadoop-root-
datanode-slave1.out
Starting secondary namenodes [master1]
master1: starting secondarynamenode, logging to /usr/local/hadoop/logs/
hadoop-root-secondarynamenode-master1.out
starting yarn daemons
starting resourcemanager, logging to /usr/local/hadoop/logs/yarn--
resourcemanager-master1.out
slave1: starting nodemanager, logging to /usr/local/hadoop/logs/yarn-root-
nodemanager-slave1.out
slave2: starting nodemanager, logging to /usr/local/hadoop/logs/yarn-root-
nodemanager-slave2.out
```

启动成功后，分别在master1、slave1和slave2节点上查看进程信息。

在master1节点上的操作命令及结果如下。

```
[root@master1 /]# jps
627 ResourceManager
453 SecondaryNameNode
889 Jps
285 NameNode
```

在slave1节点上的操作命令及结果如下。

```
[root@slave1 /]# jps
145 DataNode
233 NodeManager
350 Jps
```

在slave2节点上的操作命令及结果如下。

```
[root@slave2 /]# jps
321 Jps
116 DataNode
204 NodeManager
```

（2）集群环境的搭建

① 在master1节点上，进入安装包的存储目录，执行"tar -zxvf hbase-1.6.0-bin.tar.gz -C /usr/local"命令，解压HBase包，操作命令如下。

```
[root@master1 opt]# tar -zxvf hbase-1.6.0-bin.tar.gz -C /usr/local
```

② 将解压后的HBase包重命名为hbase，操作命令如下。

```
[root@master1 local]# mv hbase-1.6.0 hbase
```

③ 编辑~/.bashrc文件，将HBase的安装路径添加到环境变量中，操作内容如下。

```
export HBASE_HOME=/usr/local/hbase
PATH=$PATH:$HBASE_HOME/bin
```

项目 4 HBase 环境搭建与基本操作

④ 修改完成后，按"Esc"键返回命令行模式，通过输入"：wq!"保存修改并退出，随后执行"source ~/.bashrc"命令，使配置文件立即生效，操作命令如下。

```
[root@master1 local]$ source ~/.bashrc
```

⑤ 进入 hbase/conf 目录，执行"ls"命令，查看配置文件，主要关注 hbase-env.sh、hbase-site.xml 和 regionservers 文件，操作命令及结果如下。

```
[root@master1 conf]# ls
hadoop-metrics2-hbase.properties   hbase-env.sh     hbase-site.xml
log4j.properties
hbase-env.cmd                      hbase-policy.xml log4j-hbtop.properties
regionservers
```

⑥ 编辑配置文件 hbase-env.sh，找到配置文件中对应的配置项，并修改配置项内容如下。

```
export JAVA_HOME=/usr/lib/jdk1.8.0
export HBASE_CLASSPATH=/usr/local/hadoop/etc/hadoop
export HBASE_MANAGES_ZK=false
```

⑦ 编辑配置文件 hbase-site.xml，修改配置项内容如下。

```xml
<configuration>
<property>
<name>hbase.rootdir</name>
<value>hdfs://master1:9000/hbase</value>
</property>
<property>
<name>hbase.cluster.distributed</name>
<value>true</value>
</property>
<property>
<name>hbase.master</name>
<value>hdfs://master1:60000</value>
</property>
<property>
<name>hbase.zookeeper.quorum</name>
<value>master1:2181,slave1:2181,slave2:2181</value>
</property>
</configuration>
```

⑧ 编辑配置文件 regionservers，修改配置项内容如下。

```
master1
slave1
slave2
```

⑨ 将 master1 节点上的 hbase 目录复制到 slave1 和 slave2 两个节点的/usr/local 目录中，操作命令如下。

```
[root@master1 ~]# scp -r /usr/local/hbase slave1:/usr/local/
[root@master1 ~]# scp -r /usr/local/hbase slave2:/usr/local/
```

（3）启动 HBase

① 分别进入 master1、slave1 和 slave2 节点的 zookeeper/bin 目录，执行"./zkServer.sh start"命令，启动 ZooKeeper 集群。

在 master1 节点上的操作命令及结果如下。

```
[root@master1 bin]# ./zkServer.sh start
ZooKeeper JMX enabled by default
Using config: /usr/local/zookeeper/bin/../conf/zoo.cfg
Starting zookeeper ... STARTED
```

在 slave1 节点上的操作命令及结果如下。

```
[root@slave1 bin]# ./zkServer.sh start
ZooKeeper JMX enabled by default
Using config: /usr/local/zookeeper/bin/../conf/zoo.cfg
Starting zookeeper ... STARTED
```

在 slave2 节点上的操作命令及结果如下。

```
[root@slave2 bin]# ./zkServer.sh start
ZooKeeper JMX enabled by default
Using config: /usr/local/zookeeper/bin/../conf/zoo.cfg
Starting zookeeper ... STARTED
```

由于 HBase 是基于 HDFS 的，故应该先启动 Hadoop。另外，由于使用了独立的 ZooKeeper 实例，故需要先手动启动 ZooKeeper 实例，然后启动 HBase，即启动顺序是 Hadoop→ZooKeeper→HBase；如果系统使用了内置的 ZooKeeper，则启动顺序是 Hadoop→HBase。停止顺序与启动顺序正好相反。

② 分别在 master1、slave1 和 slave2 节点上执行"./zkServer.sh status"命令，查看 ZooKeeper 节点的状态。

在 master1 节点上查看节点状态的操作命令及结果如下。

```
[root@master1 bin]# ./zkServer.sh status
ZooKeeper JMX enabled by default
Using config: /usr/local/zookeeper/bin/../conf/zoo.cfg
Mode: follower
```

在 slave1 节点上查看节点状态的操作命令及结果如下。

```
[root@slave1 bin]# ./zkServer.sh status
ZooKeeper JMX enabled by default
Using config: /usr/local/zookeeper/bin/../conf/zoo.cfg
Mode: follower
```

在 slave2 节点上查看节点状态的操作命令及结果如下。

```
[root@slave2 bin]# ./zkServer.sh status
ZooKeeper JMX enabled by default
Using config: /usr/local/zookeeper/bin/../conf/zoo.cfg
Mode: leader
```

③ 在 master1 节点上进入 /home/hbase/bin 目录，执行"./start-hbase.sh"命令，启动 HBase，

项目 4　HBase 环境搭建与基本操作

操作命令及结果如下。

```
[root@master1 bin]# ./start-hbase.sh
/usr/local/hbase/conf/hbase-env.sh: line 1: !/usr/bin/env: No such file or directory
SLF4J: Class path contains multiple SLF4J bindings.
SLF4J: Found binding in [jar:file:/usr/local/hadoop/share/hadoop/common/lib/
slf4j-log4j12-1.7.10.jar!/org/slf4j/impl/StaticLoggerBinder.class]
SLF4J: Found binding in [jar:file:/usr/local/hbase/lib/client-facing-
thirdparty/slf4j-log4j12-1.7.25.jar!/org/slf4j/impl/StaticLoggerBinder.class]
SLF4J: Actual binding is of type [org.slf4j.impl.Log4jLoggerFactory]
SLF4J: Class path contains multiple SLF4J bindings.
```

④ HBase 启动成功后，分别在 master1、slave1 和 slave2 节点上执行 "jps" 命令，查看进程信息。

在 master1 节点上的操作命令及结果如下。

```
[root@master1 bin]# jps
452 SecondaryNameNode
932 QuorumPeerMain
1463 Jps
248 NameNode
1276 HRegionServer
621 ResourceManager
1135 HMaster
```

在 slave1 节点上的操作命令及结果如下。

```
[root@slave1 bin]# jps
499 HRegionServer
648 Jps
395 QuorumPeerMain
126 DataNode
239 NodeManager
```

在 slave2 节点上的操作命令及结果如下。

```
[root@slave2 bin]# jps
385 QuorumPeerMain
498 HRegionServer
116 DataNode
229 NodeManager
639 Jps
```

⑤ 在 master1 节点上执行 "./hbase shell" 命令，进入 HBase 的命令行模式，操作命令及结果如下。

```
[root@master1 bin]# ./hbase shell
SLF4J: Class path contains multiple SLF4J bindings.
SLF4J: Found binding in [jar:file:/usr/local/hadoop/share/hadoop/common/lib/
slf4j-log4j12-1.7.10.jar!/org/slf4j/impl/StaticLoggerBinder.class]
```

```
SLF4J: Found binding in [jar:file:/usr/local/hbase/lib/client-facing-
thirdparty/slf4j-log4j12-1.7.25.jar!/org/slf4j/impl/StaticLoggerBinder.class]
SLF4J: See http://www.slf4j.org/codes.html#multiple_bindings for an explanation.
SLF4J: Actual binding is of type [org.slf4j.impl.Log4jLoggerFactory]
HBase Shell
Use "help" to get list of supported commands.
Use "exit" to quit this interactive shell.
Took 0.0021 seconds
hbase(main):001:0>
```

⑥ 在 HBase Shell 中执行命令"status 'detailed'"命令，查看集群的运行状态，如图 4-3 所示。

```
hbase(main):001:0> status 'detailed'
version 1.6.0
0 regionsInTransition
active master: master1:16000 1678431792945
0 backup masters
master coprocessors: null
3 live servers
    master1:16020 1678431795609
        requestsPerSecond=0.0, numberOfOnlineRegions=1, usedHeapMB=32, maxHeapMB=1918, numberOfStores=1,
 numberOfStorefiles=1, storefileUncompressedSizeMB=0, storefileSizeMB=0, memstoreSizeMB=0, storefileInde
xSizeMB=0, readRequestsCount=0, writeRequestsCount=0, rootIndexSizeKB=0, totalStaticIndexSizeKB=0, total
StaticBloomSizeKB=0, totalCompactingKVs=0, currentCompactedKVs=0, compactionProgressPct=NaN, coprocessor
s=[]
        "user_action,,1677990108894.ac4f1d80da56e40642f535f6fb226b92."
            numberOfStores=1, numberOfStorefiles=1, storeRefCount=0, maxCompactedStoreFileRefCount=0, st
orefileUncompressedSizeMB=0, lastMajorCompactionTimestamp=0, storefileSizeMB=0, memstoreSizeMB=0, storef
ileIndexSizeMB=0, readRequestsCount=0, writeRequestsCount=0, rootIndexSizeKB=0, totalStaticIndexSizeKB=0
, totalStaticBloomSizeKB=0, totalCompactingKVs=0, currentCompactedKVs=0, compactionProgressPct=NaN, comp
leteSequenceId=-1, dataLocality=0.0
    slave1.hadoop:16020 1678431796550
        requestsPerSecond=0.0, numberOfOnlineRegions=1, usedHeapMB=39, maxHeapMB=1918, numberOfStores=1,
 numberOfStorefiles=1, storefileUncompressedSizeMB=0, storefileSizeMB=0, memstoreSizeMB=0, storefileInde
xSizeMB=0, readRequestsCount=4, writeRequestsCount=0, rootIndexSizeKB=0, totalStaticIndexSizeKB=0, total
StaticBloomSizeKB=0, totalCompactingKVs=0, currentCompactedKVs=0, compactionProgressPct=NaN, coprocessor
s=[]
        "hbase:namespace,,1677988993168.d222b7f710878ed71ac636ad037e329f."
            numberOfStores=1, numberOfStorefiles=1, storeRefCount=0, maxCompactedStoreFileRefCount=0, st
orefileUncompressedSizeMB=0, lastMajorCompactionTimestamp=0, storefileSizeMB=0, memstoreSizeMB=0, storef
ileIndexSizeMB=0, readRequestsCount=4, writeRequestsCount=0, rootIndexSizeKB=0, totalStaticIndexSizeKB=0
, totalStaticBloomSizeKB=0, totalCompactingKVs=0, currentCompactedKVs=0, compactionProgressPct=NaN, comp
leteSequenceId=-1, dataLocality=0.0
    slave2.hadoop:16020 1678431796734
        requestsPerSecond=0.0, numberOfOnlineRegions=2, usedHeapMB=45, maxHeapMB=1918, numberOfStores=3,
 numberOfStorefiles=2, storefileUncompressedSizeMB=0, storefileSizeMB=0, memstoreSizeMB=0, storefileInde
xSizeMB=0, readRequestsCount=13, writeRequestsCount=3, rootIndexSizeKB=0, totalStaticIndexSizeKB=0, tota
lStaticBloomSizeKB=0, totalCompactingKVs=0, currentCompactedKVs=0, compactionProgressPct=NaN, coprocesso
rs=[MultiRowMutationEndpoint]
        "hbase:meta,,1"
            numberOfStores=1, numberOfStorefiles=2, storeRefCount=0, maxCompactedStoreFileRefCount=0, st
orefileUncompressedSizeMB=0, lastMajorCompactionTimestamp=1677994785002, storefileSizeMB=0, memstoreSize
MB=0, storefileIndexSizeMB=0, readRequestsCount=13, writeRequestsCount=3, rootIndexSizeKB=0, totalStatic
IndexSizeKB=0, totalStaticBloomSizeKB=0, totalCompactingKVs=0, currentCompactedKVs=0, compactionProgress
Pct=NaN, completeSequenceId=-1, dataLocality=0.0
        "ip_info,,1677994346945.969d8cbac5035f8eb7f21ff1336ca866."
            numberOfStores=2, numberOfStorefiles=0, storeRefCount=0, maxCompactedStoreFileRefCount=0, st
orefileUncompressedSizeMB=0, lastMajorCompactionTimestamp=0, storefileSizeMB=0, memstoreSizeMB=0, storef
ileIndexSizeMB=0, readRequestsCount=0, writeRequestsCount=0, rootIndexSizeKB=0, totalStaticIndexSizeKB=0
, totalStaticBloomSizeKB=0, totalCompactingKVs=0, currentCompactedKVs=0, compactionProgressPct=NaN, comp
leteSequenceId=-1, dataLocality=0.0
0 dead servers
```

图 4-3 查看集群的运行状态

⑦ 在 Windows 浏览器地址栏中输入"http://宿主机 IP 地址:16010 对应的映射端口:/master-status"，查看 HBase 的管理界面，如图 4-4 所示。

项目 4　HBase 环境搭建与基本操作

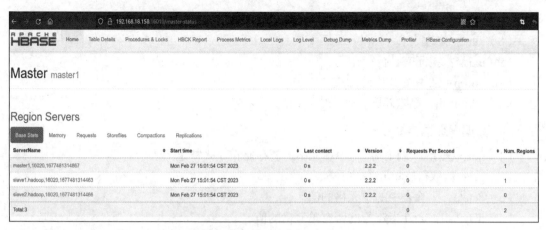

图 4-4　查看 HBase 的管理界面

任务 4.2　HBase Shell 操作

📖 任务描述

了解 HBase Shell 的相关知识，学习 HBase Shell 的常用操作。

📖 任务目标

（1）熟悉 HBase Shell 的相关知识。
（2）学会 HBase Shell 的常用操作。

📖 知识准备

HBase 包含可以与 HBase 进行通信的 Shell。HBase Shell 支持的命令如下。

1. 通用命令

（1）status：提供 HBase 的状态信息，如服务器的数量。
（2）version：提供正在使用的 HBase 的版本信息。
（3）table_help：查看关于表操作相关命令的帮助信息。
（4）whoami：提供有关用户的信息。

2. 数据定义语言命令

（1）create：创建表。
（2）list：列出 HBase 的所有表。
（3）disable：禁用表。
（4）is_disabled：验证表是否被禁用。
（5）enable：启用表。
（6）is_enabled：验证表是否已启用。

119

（7）describe：提供表的描述信息。
（8）alter：修改表。
（9）exists：验证表是否存在。
（10）drop：从 HBase 中删除表。
（11）drop_all：删除匹配在命令中给出的正则表达式的表。

3. 数据操纵语言命令

（1）put：把指定列和指定行中单元格的值存储在一张特定的表中。
（2）get：获取行或单元格的内容。
（3）delete：删除表中的单元格的值。
（4）deleteall：删除指定行的所有单元格的值。
（5）scan：扫描并返回表数据。
（6）count：计数并返回表中行的数目。
（7）truncate：禁用、删除和重新创建一张指定的表。

任务实施

1. 启动 HBase

（1）配置 Hadoop 分布式环境

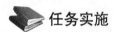

微课 4-4　HBase Shell 操作

在 master1、slave1 和 slave2 节点上配置 Hadoop 完全分布式环境。在 master1 节点上执行 "start-all.sh" 命令，启动 Hadoop 集群，操作命令及结果如下。

```
[root@master1 /]# start-all.sh
This script is Deprecated. Instead use start-dfs.sh and start-yarn.sh
Starting namenodes on [master1]
master1: starting namenode, logging to /usr/local/hadoop/logs/hadoop-root-namenode-master1.out
slave2: starting datanode, logging to /usr/local/hadoop/logs/hadoop-root-datanode-slave2.out
slave1: starting datanode, logging to /usr/local/hadoop/logs/hadoop-root-datanode-slave1.out
```

Hadoop 集群启动成功后，在 master1 节点上执行 "jps" 命令，操作命令及结果如下。

```
[root@master1 /]# jps
898 Jps
452 SecondaryNameNode
248 NameNode
621 ResourceManager
```

在 slave1 节点上执行 "jps" 命令，操作命令及结果如下。

```
[root@slave1 /]# jps
371 Jps
126 DataNode
239 NodeManager
```

项目 4　HBase 环境搭建与基本操作

在 slave2 节点上执行 "jps" 命令，操作命令及结果如下。

```
[root@slave2 /]# jps
116 DataNode
229 NodeManager
361 Jps
```

（2）启动 ZooKeeper 集群

分别进入 master1、slave1 和 slave2 节点的 zookeeper/bin 目录，执行 "./zkServer.sh start" 命令启动 ZooKeeper 集群。

在 master1 节点上的操作命令及结果如下。

```
[root@master1 bin]# ./zkServer.sh start
ZooKeeper JMX enabled by default
Using config: /usr/local/zookeeper/bin/../conf/zoo.cfg
Starting zookeeper ... STARTED
```

在 slave1 节点上的操作命令及结果如下。

```
[root@slave1 bin]# ./zkServer.sh start
ZooKeeper JMX enabled by default
Using config: /usr/local/zookeeper/bin/../conf/zoo.cfg
Starting zookeeper ... STARTED
```

在 slave2 节点上的操作命令及结果如下。

```
[root@slave2 bin]# ./zkServer.sh start
ZooKeeper JMX enabled by default
Using config: /usr/local/zookeeper/bin/../conf/zoo.cfg
Starting zookeeper ... STARTED
```

分别在 master1、slave1 和 slave2 节点上执行 "./zkServer.sh status" 命令，查看节点的状态。

在 master1 节点上查看节点状态的操作命令及结果如下。

```
[root@master1 bin]# ./zkServer.sh status
ZooKeeper JMX enabled by default
Using config: /usr/local/zookeeper/bin/../conf/zoo.cfg
Mode: follower
```

在 slave1 节点上查看节点状态的操作命令及结果如下。

```
[root@slave1 bin]# ./zkServer.sh status
ZooKeeper JMX enabled by default
Using config: /usr/local/zookeeper/bin/../conf/zoo.cfg
Mode: follower
```

在 slave2 节点上查看节点状态的操作命令及结果如下。

```
[root@slave2 bin]# ./zkServer.sh status
ZooKeeper JMX enabled by default
Using config: /usr/local/zookeeper/bin/../conf/zoo.cfg
Mode: leader
```

（3）启动 HBase 并查看进程信息

进入 master1 节点的 /home/hbase/bin 目录，执行 "./start-hbase.sh" 命令，启动 HBase，操作命令及结果如下。

```
[root@master1 bin]# ./start-hbase.sh
SLF4J: Class path contains multiple SLF4J bindings.
SLF4J: Found binding in [jar:file:/usr/local/hadoop/share/hadoop/common/lib/
slf4j-log4j12-1.7.10.jar!/org/slf4j/impl/StaticLoggerBinder.class]
SLF4J: Found binding in [jar:file:/usr/local/hbase/lib/client-facing-thirdparty/
slf4j-log4j12-1.7.25.jar!/org/slf4j/impl/StaticLoggerBinder.class]
SLF4J: Actual binding is of type [org.slf4j.impl.Log4jLoggerFactory]
running master, logging to /usr/local/hbase/logs/hbase--master-master1.out
SLF4J: Class path contains multiple SLF4J bindings.
```

HBase 启动成功后，分别在 master1、slave1 和 slave2 节点上执行 "jps" 命令，查看 HBase 进程信息。

在 master1 节点上的操作命令及结果如下。

```
[root@master1 bin]# jps
452 SecondaryNameNode
932 QuorumPeerMain
1463 Jps
248 NameNode
1276 HRegionServer
621 ResourceManager
1135 HMaster
```

在 slave1 节点上的操作命令及结果如下。

```
[root@slave1 bin]# jps
499 HRegionServer
648 Jps
395 QuorumPeerMain
126 DataNode
239 NodeManager
```

在 slave2 节点上的操作命令及结果如下。

```
[root@slave2 bin]# jps
385 QuorumPeerMain
498 HRegionServer
116 DataNode
229 NodeManager
639 Jps
```

2. HBase Shell 操作

（1）创建及查看表

① 在 master1 节点上执行 "./hbase shell" 命令，进入 HBase 命令行模式，操作命令及结果如下。

项目 4　HBase 环境搭建与基本操作

```
[root@master1 bin]# ./hbase shell
SLF4J: Class path contains multiple SLF4J bindings.
SLF4J: Found binding in [jar:file:/usr/local/hadoop/share/hadoop/common/lib/
slf4j-log4j12-1.7.10.jar!/org/slf4j/impl/StaticLoggerBinder.class]
SLF4J: Found binding in [jar:file:/usr/local/hbase/lib/client-facing-
thirdparty/slf4j-log4j12-1.7.25.jar!/org/slf4j/impl/StaticLoggerBinder.class]
SLF4J: Actual binding is of type [org.slf4j.impl.Log4jLoggerFactory]
HBase Shell
Use "help" to get list of supported commands.
Use "exit" to quit this interactive shell.
Took 0.0021 seconds
hbase(main):001:0>
```

　　② 创建 student 表，其属性有 name、sex、age、dept、course，因为 HBase 的表中会有一个系统默认的属性作为行键，所以无须自行创建行键，操作命令及结果如下。

```
hbase(main):001:0> create 'student','name','sex','age','dept','course'
Created table student
Took 1.7020 seconds
=> Hbase::Table - student
```

　　③ 创建好 student 表后，执行"describe 'student'"命令，查看 student 表的基本信息，操作命令及结果如下。

```
hbase(main):002:0> describe 'student'
Table student is ENABLED
student
COLUMN FAMILIES DESCRIPTION
{NAME => 'age', VERSIONS => '1', EVICT_BLOCKS_ON_CLOSE => 'false',
NEW_VERSION_BEHAVIOR => 'false', KEEP_DELETED_CELLS => 'FALSE', CACHE_DATA_ON
_WRITE => 'false', DATA_BLOCK_ENCODING => 'NONE', TTL => 'FOREVER', MIN_VERSIONS
=> '0', REPLICATION_SCOPE => '0', BLOOMFILTER => 'ROW', CACHE_I
INDEX_ON_WRITE => 'false', IN_MEMORY => 'false', CACHE_BLOOMS_ON_WRITE => 'false',
PREFETCH_BLOCKS_ON_OPEN => 'false', COMPRESSION => 'NONE', BLO
CKCACHE => 'true', BLOCKSIZE => '65536'}

#省略部分信息
5 row(s)

QUOTAS
0 row(s)
Took 0.2088 seconds
```

　　④ 执行"list"命令，查看所有表，操作命令及结果如下。

```
hbase(main):003:0> list
TABLE
student
```

```
1 row(s)
Took 0.0249 seconds
```

（2）添加表数据

在添加数据时，HBase 会自动为添加的数据增加一个时间戳，故在需要修改数据时，只需直接添加数据，HBase 便会生成一个新版本，从而完成"改"操作，旧的版本依旧保留，系统会定时回收垃圾数据，只留下最新的几个版本，保存的版本数可以在创建表的时候指定。HBase 中使用"put"命令添加数据，一次只能为一张表的一行数据的一个列（即一个单元格）添加数据，所以直接使用 Shell 命令插入数据的效率很低，在实际应用中，一般利用编程插入数据。

例如，当执行"put 'student','95001','name','LiMing'"命令时，表示为 student 表添加了学号为 95001、名字为 LiMing 的一行数据，其行键为 95001，操作命令及结果如下。

```
hbase(main):004:0> put 'student','95001','name','LiMing'
Took 0.0865 seconds
```

当执行"put 'student','95001','course:math','80'"命令时，表示为 95001 行下的 course 列族的 math 列添加数据，操作命令及结果如下。

```
hbase(main):005:0>
hbase(main):005:0> put 'student','95001','course:math','80'
Took 0.0158 seconds
```

（3）删除表数据

在 HBase 中使用"delete"及"deleteall"命令可进行删除数据操作。"delete"命令用于删除一个数据，是"put"的反向操作；"deleteall"命令用于删除一行数据。

删除 student 表中 95001 行下的 sex 列的数据，操作命令及结果如下。

```
hbase(main):006:0> delete 'student','95001','sex'
Took 0.0095 seconds
hbase(main):007:0> get 'student','95001'
COLUMN                  CELL
 course:math
timestamp=1677484606134, value=80
 name:
timestamp=1677484558867, value=LiMing
1 row(s)
Took 0.0291 seconds
```

删除 student 表中的 95001 行的全部数据，操作命令及结果如下。

```
hbase(main):008:0> deleteall 'student','95001'
Took 0.0139 seconds
hbase(main):009:0> scan 'student'
ROW                                    COLUMN+CELL
0 row(s)
Took 0.0169 seconds
```

（4）查看表数据

HBase 中有两个用于查看数据的命令："get"命令用于查看表的某一行的数据；"scan"

命令用于查看某张表的全部数据。

查看 student 表中的 95001 行的数据，操作命令及结果如下。

```
hbase(main):010:0> get 'student','95001'
COLUMN                CELL
0 row(s)
Took 0.0062 seconds
hbase(main):011:0> scan 'student'
ROW                                             COLUMN+CELL
0 row(s)
Took 0.0086 seconds
```

（5）删除表

当需要删除表时，要先使此表不可用（使用"disable"命令实现），再进行删除表操作（使用"drop"命令实现），操作命令及结果如下。

```
hbase(main):012:0> disable 'student'
Took 2.2791 seconds
hbase(main):013:0> drop 'student'
Took 0.4499 seconds
hbase(main):014:0> list
TABLE
0 row(s)
Took 0.0074 seconds
=> [
```

（6）查询表版本

当需要查询表的历史版本时，操作步骤如下。

① 在创建表时指定保存的版本数（假设指定为 5），操作命令及结果如下。

```
hbase(main):016:0> create 'teacher',{NAME=>'username',VERSIONS=>5}
Created table teacher
Took 0.7335 seconds
=> Hbase::Table - teacher
```

② 插入数据并更新数据，使其产生历史版本数据。注意：这里插入数据和更新数据都使用了"put"命令，操作命令及结果如下。

```
hbase(main):017:0> put 'teacher','91001','username','Mary'
Took 0.0714 seconds
hbase(main):018:0> put 'teacher','91001','username','Mary1'
Took 0.0061 seconds
hbase(main):019:0> put 'teacher','91001','username','Mary2'
Took 0.0046 seconds
hbase(main):020:0> put 'teacher','91001','username','Mary3'
Took 0.0043 seconds
hbase(main):021:0> put 'teacher','91001','username','Mary4'
Took 0.0155 seconds
```

```
hbase(main):022:0> put 'teacher','91001','username','Mary5'
Took 0.0066 seconds
hbase(main):023:0>
```

③ 查询时，指定查询的历史版本数（有效取值为1～5），若未指定，则默认查询出最新的数据，操作命令及结果如下。

```
hbase(main):023:0> get 'teacher','91001',{COLUMN=>'username',VERSIONS=>4}
COLUMN                                              CELL
 username:                                          timestamp=1677485193849,
value=Mary5
 username:                                          timestamp=1677485191859,
value=Mary4
 username:                                          timestamp=1677485187591,
value=Mary3
 username:                                          timestamp=1677485185178,
value=Mary2
1 row(s)
Took 0.5959 seconds
hbase(main):024:0> get 'teacher','91001',{COLUMN=>'username',VERSIONS=>5}
COLUMN                                              CELL
 username:                                          timestamp=1677485193849,
value=Mary5
 username:                                          timestamp=1677485191859,
value=Mary4
 username:                                          timestamp=1677485187591,
value=Mary3
 username:                                          timestamp=1677485185178,
value=Mary2
 username:                                          timestamp=1677485181651,
value=Mary1
1 row(s)
Took 0.0080 seconds
hbase(main):025:0>
```

④ 执行"exit"命令，退出HBase数据库。注意：这里退出HBase数据库是指退出对数据库表的操作，而不是停止HBase数据库的后台运行，操作命令及结果如下。

```
hbase(main):025:0> exit
[root@master1 bin]#
```

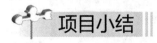项目小结

本项目主要介绍了HBase的产生背景、特点、数据结构、存储机制、架构等，并重点介

项目 4　HBase 环境搭建与基本操作

绍了 3 种类型的服务器——HRegion Server、HMaster 和 ZooKeeper 的功能，以及 HBase Shell 操作，完成了 HBase 的安装与配置。通过对本项目的学习，读者能够对 HBase 技术和应用等建立清晰的认识，为在大数据应用场景下，利用 HBase 实现数据的实时存储奠定良好基础。

课后练习

1. 请描述 HBase 的读写流程。
2. 如何提高 HBase 客户端的读写性能？
3. 请描述 HBase 实时查询的原理。
4. HBase 如何导入数据？
5. 请描述 HBase 的存储结构。
6. HBase 适用于怎样的情景？

项目 5 Hadoop 生态组件的安装与使用

学习目标

【知识目标】
识记 Hadoop 生态组件（Sqoop、Pig、Flume、Kafka、Flink）的作用。
领会 Hadoop 各组件的功能与联系。

【素质目标】
具有严谨细致的工作态度和工作作风。
具有良好的团队协作意识和业务沟通能力。
具有良好的表达能力和文档查阅能力。
具有规范的编程意识和较好的数据洞察力。

【技能目标】
学会 Hadoop 生态组件（Sqoop、Pig、Flume、Kafka、Flink）的安装。
学会 Hadoop 生态组件（Sqoop、Pig、Flume、Kafka、Flink）的使用。

项目描述

Hadoop 包含很多组件，其中有两个核心组件——被称为 Hadoop 分布式文件系统的 HDFS，以及被称为分布式计算框架的 MapReduce。有一些支持项目充分利用了 HDFS 和 MapReduce。在做大数据分析业务时，具体使用哪种组件需要视业务逻辑需求而定。

本项目主要介绍 Hadoop 生态组件 Sqoop、Pig、Flume、Kafka、Flink 的安装配置及简单使用。

任务 5.1 Sqoop 的安装与使用

任务描述

（1）学习 Sqoop 的相关知识，熟悉 Sqoop 的作用、完成 Sqoop 的安装与配置等。
（2）使用 Sqoop 完成 MySQL 和 HDFS 之间的数据互导。

任务目标

（1）学会 Sqoop 的安装与配置。
（2）学会使用 Sqoop 完成 MySQL 和 HDFS 之间的数据互导。

项目 5　Hadoop 生态组件的安装与使用

知识准备

Sqoop（SQL-to-Hadoop）是一种开源工具，主要用于在 Hadoop 与传统的数据库（如 MySQL、Oracle 等）之间进行数据传递，它可以将一个关系数据库（如 MySQL、Oracle、PostgreSQL 等）中的数据导入 Hadoop 的 HDFS 中，也可以将 HDFS 的数据导入关系数据库中。

对于某些非关系数据库（如 NoSQL 等），Sqoop 也提供了连接器。Sqoop 类似于其他 ETL（Extract-Transform-Load，用来描述将数据从源端经过抽取、转换、装载至目的端的过程）工具，使用元数据模型来判断数据类型，并在数据从数据源转移到 Hadoop 时保证类型安全，确保数据完整性。Sqoop 专为大数据批量传输而设计，能够通过分割数据集并创建 Hadoop 任务来处理每个区块的数据。

Sqoop 项目开始于 2009 年，其最早作为 Hadoop 的一个第三方模块而存在，后来为了让使用者快速部署，也为了使程序开发人员更快速地进行迭代开发，Sqoop 成为一个独立的 Apache 项目。

1. Sqoop 的核心功能

Sqoop 的核心功能包含以下 2 项。

（1）导入数据

从 MySQL、Oracle 导入数据到 Hadoop 的 HDFS、Hive、HBase 等数据存储系统中。

（2）导出数据

从 Hadoop 的 HDFS 中导出数据到关系数据库（如 MySQL 等）中。Sqoop 本质上是一种命令行工具。

2. Sqoop 中 import 命令的使用

（1）默认情况下，通过使用 "import" 命令可将数据导入 HDFS 中，操作命令如下。

```
$ /bin/sqoop import \
--connect jdbc:mysql://hostname:3306/mydb \
--username root \
--password root \
--table mytable
```

（2）指定目录和 Mapper（Hadoop 提供的抽象类，支持继承和实现其中相关接口函数）个数，并将数据导入 HDFS 中。

① 创建目录的操作命令如下。

```
${HADOOP_HOME}/bin/hdfs dfs -mkdir -p /user/sqoop/
```

② 设置 Mapper 个数为 1，指定目标目录为/user/hive/warehouse/my_user，如果目标目录已经存在，则先删除原目录，再创建新目录，操作命令如下。

```
$ bin/sqoop import \
--connect jdbc:mysql://blue01.mydomain:3306/mydb \
--username root \
--password root \
```

```
--table my_user \
--target-dir /user/hive/warehouse/my_user \
--delete-target-dir \
--num-mappers 1 \
--fields-terminated-by "\t" \
--columns id,passwd \
--where "id<=3"
```

(3)将增量数据导入 HDFS 中,可以通过对"check-column""incremental""last-value"这 3 个参数进行设置来实现,操作命令如下。

```
$bin/sqoop import \
--connect jdbc:mysql://hostname:3306/mydb \
--username root \
--password root \
--table mytable \
--num-mappers 1 \
--target-dir /user/sqoop/ \
--fields-terminated-by "\t" \
--check-column id \
--incremental append \
--last-value 4        //表示从第 5 位开始导入
```

(4)指定文件存储格式并将数据导入 HDFS 中。默认情况下,将数据导入 HDFS 中时,文件存储格式为 textfile,可以通过对属性进行设置以指定文件存储格式为 parquet,操作命令如下。

```
$bin/sqoop import \
--connect jdbc:mysql://hostname:3306/mydb \
--username root \
--password root \
--table mytable \
--num-mappers 1 \
--target-dir /user/sqoop/ \
--fields-terminated-by "\t" \
--as-parquetfile
```

(5)指定压缩格式并将数据导入 HDFS 中。默认情况下,导入时是不压缩的,可以通过对属性"compress""compression-codec"进行设置来指定压缩格式,操作命令如下。

```
$bin/sqoop import \
--connect jdbc:mysql://hostname:3306/mydb \
--username root \
--password root \
--table mytable \
--num-mappers 1 \
--target-dir /user/sqoop/ \
```

```
--fields-terminated-by "\t" \
--compress \
--compression-codec org.apache.hadoop.io.compress.SnappyCodec
```

（6）将查询结果导入 HDFS 中时，必须在 where 子句中包含 "$CONDITIONS"，操作命令如下。

```
$ bin/sqoop import \
--connect jdbc:mysql://hostname:3306/mydb \
--username root \
--password root \
--target-dir /user/hive/warehouse/mydb.db/mytable \
--delete-target-dir \
--num-mappers 1 \
--fields-terminated-by "\t" \
--query 'select id,account from my_user where id>=3 and $CONDITIONS'
```

（7）将数据导入 Hive 中，操作命令如下。

```
$ bin/sqoop import \
--connect jdbc:mysql://hostname:3306/mydb \
--username root \
--password root \
--table mytable \
--num-mappers 1 \
--hive-import \
--hive-database mydb \
--hive-table mytable \
--fields-terminated-by "\t" \
--delete-target-dir \
--hive-overwrite
```

3. Sqoop 中 export 命令的使用

（1）这里以数据导入为例进行说明，操作命令如下。

① 将数据导入 HDFS 中。

```
$ bin/sqoop export \
--connect jdbc:mysql://hostname:3306/mydb
--username root
--password root
--table mytable
--num-mappers 1
--export-dir /user/hive/warehouse/mydb.db/mytable
--input-fields-terminated-by "\t"
```

② 执行脚本。

```
$ bin/sqoop --options-file ***.opt
```

（2）从 Hive 或者 HDFS 中导出数据到 MySQL 中，操作命令如下。

```
$ bin/sqoop export \
--connect jdbc:mysql://hostname:3306/mydb \
--username root \
--password root \
--table mytable \
--num-mappers 1 \
--export-dir /user/hive/warehouse/mydb.db/mytable \
--input-fields-terminated-by "\t"
```

任务实施

1. 安装与配置 Sqoop

（1）安装与配置所需的软件

① 切换到安装包的存储目录，执行"tar -zxvf sqoop-1.4.7.bin_hadoop-2.6.0.tar.gz -C /usr/local"命令，将 Sqoop 安装包解压到指定目录中，操作命令如下。

微课 5-1　Sqoop 的安装与使用

```
[root@master1 opt]# tar -zxvf sqoop-1.4.7.bin_hadoop-2.6.0.tar.gz -C /usr/local
```

② 将解压后的 Sqoop 安装包重命名为"sqoop"，操作命令如下。

```
[root@master1 local]# mv sqoop-1.4.7.bin_hadoop-2.6.0 sqoop
```

如果在普通用户模式下，则需要额外执行"sudo chown -R hadoop:hadoop sqoop/"命令，修改 sqoop 目录的权限。

（2）配置 sqoop-env.sh 文件

① 执行"cat sqoop-env-template.sh >>sqoop-env.sh"命令，复制/usr/local/sqoop/conf 目录中的配置文件 sqoop-env-template.sh，并将其重命名为 sqoop-env.sh，操作命令如下。

```
[root@master1 conf]# cat sqoop-env-template.sh >>sqoop-env.sh
```

② 编辑 sqoop-env.sh 文件，分别将 Hadoop、HBase、Hive 等的安装目录添加到文件中，编辑内容如下。

```
#Set the path to where bin/hadoop is available
export HADOOP_COMMON_HOME=/usr/local/hadoop
#Set the path to where hadoop-*-core.jar is available
export HADOOP_MAPRED_HOME=/usr/lcoal/hadoop
#set the path to where bin/hbase is available
export HBASE_HOME=/usr/local/hbase
#Set the path to where bin/hive is available
export HIVE_HOME=/simple/hive1.2.1
```

（3）配置环境变量

① 编辑 ~/.bashrc 文件，将 Sqoop 的安装路径添加到环境变量中，编辑内容如下。

```
export SQOOP_HOME=/usr/local/sqoop
PATH=$PATH:$SQOOP_HOME/bin
```

项目 5 Hadoop 生态组件的安装与使用

② 修改完成后，按"Esc"键返回命令行模式，通过输入":wq!"保存修改并退出，随后执行"source ~/.bashrc"命令，使配置文件立即生效，操作命令如下。

```
[root@master local]$ source ~/.bashrc
```

（4）配置 MySQL 连接

执行"cp mysql-connector-java-5.1.40-bin.jar /usr/local/sqoop/lib/"命令，将数据库连接需要的 JAR 包添加到 /usr/local/sqoop/lib 目录下，操作命令如下。

```
[root@master1 mysql-connector-java-5.1.40]# cp mysql-connector-java-5.1.40-bin.jar /usr/local/sqoop/lib/
```

（5）测试 Sqoop 与 MySQL 之间的连接

① 执行"service mysql start"命令，启动 MySQL 服务，操作命令及结果如下。

```
[root@master1 mysql-connector-java-5.1.40]# service mysql start
Starting MySQL SUCCESS!
```

② 执行"sqoop list-databases --connect jdbc:mysql://localhost:3306 --username root --password 123456"命令，测试 Sqoop 与 MySQL 之间的连接是否成功，如果可以看到 MySQL 数据库中的列表，则表示 Sqoop 与 MySQL 连接成功。若连接时提示与安全套接字层（Secure Socket Layer，SSL）相关的信息，则可以增加 useSSL=false 参数来进行配置，操作命令及结果如下。

```
[root@master1 bin]# ./sqoop list-databases --connect
jdbc:mysql://localhost:3306/?useSSL=false --username root --password 123456
23/03/02 11:15:31 INFO sqoop.Sqoop: Running Sqoop version: 1.4.7
23/03/02 11:15:31 WARN tool.BaseSqoopTool: Setting your password on the
command-line is insecure. Consider using -P instead.
23/03/02 11:15:31 INFO manager.MySQLManager: Preparing to use a MySQL streaming
resultset.
information_schema
info_station
information_DB
myhive
mysql
performance_schema
sys
```

2. 使用 Sqoop 完成 MySQL 和 HDFS 之间的数据互导

（1）将准备好的测试数据上传到 MySQL 中

① 登录 MySQL，操作命令及结果如下。

```
[root@master1 opt]# mysql -u root -p
Enter password:
Welcome to the MySQL monitor.  Commands end with ; or \g.
Your MySQL connection id is 2
Server version: 5.7.23-log MySQL Community Server (GPL)
Copyright (c) 2000, 2018, Oracle and/or its affiliates. All rights reserved.
Oracle is a registered trademark of Oracle Corporation and/or its
```

```
affiliates. Other names may be trademarks of their respective
owners.
Type 'help;' or '\h' for help. Type '\c' to clear the current input statement.
mysql>
```

② 在测试数据库 test 中创建表 test1，用于存放本地测试数据，操作命令及结果如下。

```
mysql> create database test \
    -> character set utf8;
Query OK, 1 row affected (0.00 sec)
mysql> use test
Database changed
mysql> create table test1(
    -> ip varchar(100) not null,
    -> time varchar(100) not null,
    -> url varchar(255) not null);
Query OK, 0 rows affected (0.01 sec)
```

③ 将本地的测试数据 test.txt 文件上传到 test1 表中，注意分隔符，此处分隔符是","，具体分隔符以实际操作为准，操作命令及结果如下。

```
mysql> load data local infile '/opt/test.txt' into table test1 fields terminated by ',';
Query OK, 30 rows affected, 60 warnings (0.00 sec)
Records: 30  Deleted: 0  Skipped: 0  Warnings: 60
```

④ 上传完成后，查看 test1 表中的前 10 条数据，如图 5-1 所示。

```
mysql> select * from test1 limit 10;
+----------------+----------------+----------------------------------------+
| ip             | time           | url                                    |
+----------------+----------------+----------------------------------------+
| 27.19.74.143   | 2015/3/30 17:38 | /static/image/common/faq.gif          |
| 110.52.250.126 | 2015/3/30 17:38 | /data/cache/style_1_widthauto.css?y7a |
| 27.19.74.143   | 2015/3/30 17:38 | /static/image/common/hot_1.gif        |
| 27.19.74.143   | 2015/3/30 17:38 | /static/image/common/hot_2.gif        |
| 27.19.74.143   | 2015/3/30 17:38 | /static/image/filetype/common.gif     |
| 110.52.250.126 | 2015/3/30 17:38 | /source/plugin/wsh_wx/img/wsh_zk.css  |
| 110.52.250.126 | 2015/3/30 17:38 | /data/cache/style_1_forum_index.css?y7a |
| 110.52.250.126 | 2015/3/30 17:38 | /source/plugin/wsh_wx/img/wx_jqr.gif  |
| 27.19.74.143   | 2015/3/30 17:38 | /static/image/common/recommend_1.gif  |
| 110.52.250.126 | 2015/3/30 17:38 | /static/image/common/logo.png         |
+----------------+----------------+----------------------------------------+
10 rows in set (0.00 sec)
```

图 5-1　查看 test1 表中的前 10 条数据

（2）上传数据到 HDFS 中

① 将 test1 中的数据上传到 HDFS 中，操作命令如下。

```
[root@master1 bin]# ./sqoop import --connect jdbc:mysql://localhost:3306/test?useSSL=false --username root --password 123456 --table test1 -m 1
```

② 执行完上述操作后，执行 "hdfsdfs -text /user/root/test1/ part-m-00000" 命令，在 HDFS 中查看导入的数据，如图 5-2 所示。

项目 5　Hadoop 生态组件的安装与使用

```
[root@master1 bin]# hdfs dfs -text /user/root/test1/ part-m-00000
text: '/user/root/test1': Is a directory
text: `part-m-00000': No such file or directory
[root@master1 bin]# hdfs dfs -text /user/root/test1/part-m-00000
27.19.74.143,2015/3/30 17:38,/static/image/common/faq.gif
110.52.250.126,2015/3/30 17:38,/data/cache/style_1_widthauto.css?y7a
27.19.74.143,2015/3/30 17:38,/static/image/common/hot_1.gif
27.19.74.143,2015/3/30 17:38,/static/image/common/hot_2.gif
27.19.74.143,2015/3/30 17:38,/static/image/filetype/common.gif
110.52.250.126,2015/3/30 17:38,/source/plugin/wsh_wx/img/wsh_zk.css
110.52.250.126,2015/3/30 17:38,/data/cache/style_1_forum_index.css?y7a
110.52.250.126,2015/3/30 17:38,/source/plugin/wsh_wx/img/wx_jqr.gif
27.19.74.143,2015/3/30 17:38,/static/image/common/recommend_1.gif
110.52.250.126,2015/3/30 17:38,/static/image/common/logo.png
27.19.74.143,2015/3/30 17:38,/data/attachment/common/c8/common_2_verify_icon.png
110.52.250.126,2015/3/30 17:38,/static/js/logging.js?y7a
8.35.201.144,2015/3/30 17:38,/uc_server/avatar.php?uid=29331&size=middle
27.19.74.143,2015/3/30 17:38,/data/cache/common_smilies_var.js?y7a
27.19.74.143,2015/3/30 17:38,/static/image/common/pn.png
27.19.74.143,2015/3/30 17:38,/static/image/common/swfupload.swf?preventswfcaching=1369906718144
27.19.74.143,2015/3/30 17:38,/static/image/editor/editor.gif
8.35.201.165,2015/3/30 17:38,/uc_server/data/avatar/000/35/94/42_avatar_middle.jpg
8.35.201.164,2015/3/30 17:38,/uc_server/data/avatar/000/03/13/51_avatar_middle.jpg
8.35.201.163,2015/3/30 17:38,/uc_server/data/avatar/000/04/87/94_avatar_middle.jpg
8.35.201.165,2015/3/30 17:38,/uc_server/data/avatar/000/01/01/03_avatar_middle.jpg
8.35.201.160,2015/3/30 17:38,/uc_server/data/avatar/000/04/12/85_avatar_middle.jpg
8.35.201.164,2015/3/30 17:38,/uc_server/avatar.php?uid=53635&size=middle
8.35.201.163,2015/3/30 17:38,/static/image/common/arw_r.gif
8.35.201.166,2015/3/30 17:38,/static/image/common/px.png
8.35.201.144,2015/3/30 17:38,/static/image/common/pmto.gif
8.35.201.161,2015/3/30 17:38,/static/image/common/search.png
8.35.201.163,2015/3/30 17:38,/uc_server/avatar.php?uid=57232&size=middle
8.35.201.164,2015/3/30 17:38,/uc_server/data/avatar/000/05/83/35_avatar_middle.jpg
8.35.201.160,2015/3/30 17:38,/uc_server/data/avatar/000/01/54/22_avatar_middle.jpg
```

图 5-2　在 HDFS 中查看导入的数据

（3）将数据从 HDFS 导入 MySQL 中

在将数据从 HDFS 中导出前需要先创建导出表的结构，如果导出表在数据表中不存在，则系统会报错；如果重复多次导出数据，则表中的数据会重复。

① 准备数据表。在 test 数据库中创建表 test2，可以直接复制 test1 表的结构，操作命令及结果如下。

```
mysql> create table test2 as select * from test1 where 1=2;
Query OK, 0 rows affected (0.00 sec)
Records: 0  Duplicates: 0  Warnings: 0
```

② 使用 Sqoop 将 HDFS 中的数据导入 MySQL 的 test2 表中，操作命令如下。

```
[root@master1 bin]# sqoop export --connect jdbc:mysql://master1:3306/test
--username root --password 123456 --table test2  --export-dir hdfs:///user/root/test1/part-m-00000 -m 1
```

③ 在 MySQL 中，执行 "select * from test2" 命令查看 test2 表中的数据，验证导入数据的正确性，操作命令及部分结果如下。

```
mysql>select * from test2
......
| 8.35.201.163    | 2015/3/30 17:38 | /uc_server/avatar.php?uid=57232&size=middle |
| 8.35.201.164    | 2015/3/30 17:38 | /uc_server/data/avatar/000/05/83/35_avatar_middle.jpg |
| 8.35.201.160    | 2015/3/30 17:38 | /uc_server/data/avatar/000/01/54/22_avatar_middle.jpg |
+-----------------+-----------------+-------------------------------------+
30 rows in set (0.00 sec)
```

任务 5.2　Pig 的安装与使用

任务描述

（1）学习 Pig 的相关知识、熟悉 Pig 的作用、完成 Pig 的安装与配置等。
（2）使用 Pig 完成简单的数据分析。

任务目标

（1）学会 Pig 的安装与配置。
（2）学会使用 Pig 完成简单的数据分析。

知识准备

1. Apache Pig 概述

Apache Pig 是 MapReduce 的一个抽象。它是一种工具，用于分析较大的数据集，并将数据集表示为数据流。Pig 通常与 Hadoop 一起使用，可以使用 Pig 在 Hadoop 中执行所有的数据处理操作。

当要编写数据分析程序时，Pig 中提供了一种名为 Pig Latin 的高级语言。该语言提供了各种运算符，开发人员可以利用其开发用于读取、写入和处理数据的程序。

要想使用 Pig 分析数据，开发人员需要使用 Pig Latin 语言编写脚本。在 Pig 内部，所有脚本都被转换为 Map 和 Reduce 任务。Pig 的工作原理如图 5-3 所示。

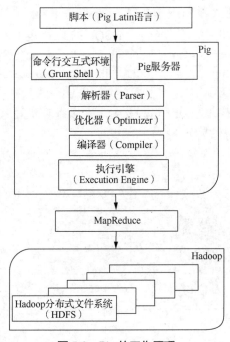

图 5-3　Pig 的工作原理

2. Pig Latin 语句基础

在使用 Pig Latin 语言处理数据时，语句是基本结构，每个语句以分号（;）结尾。Pig Latin 提供的运算符可通过语句执行各种操作。除了 LOAD 和 STORE 语句之外，在执行其他操作时，Pig Latin 语句采用关系作为输入，并产生另一个关系作为输出。只要在 Shell 中输入 LOAD 语句，就会执行语义检查。要想查看模式的内容，需要使用 Dump 运算符。只有在执行 Dump 操作后，才会执行将数据加载到文件系统中的 MapReduce 任务。

Pig Latin 的数据类型及其说明如表 5-1 所示。

表 5-1 Pig Latin 的数据类型及其说明

数据类型	说明
int	表示有符号的 32 位整数，如 8
long	表示有符号的 64 位整数，如 5L
float	表示有符号的 32 位浮点数，如 5.5F
double	表示有符号的 64 位浮点数，如 10.5
chararray	表示 UTF-8 格式的字符数组，如'w3cschool'
bytearray	表示字节数组
boolean	表示布尔值，如 true/false
datetime	表示日期时间，如 1970-01-01T00:00:00.000 + 00:00
biginteger	表示 Java BigInteger，如 60708090709
bigdecimal	表示 Java BigDecimal，如 185.98376256272893883
tuple	元组是有序的字段集合，如(tony,30)
bag	包是元组的集合，如{(tony,30),(tom,45)}
map	映射是一组 Key-Value 对，如['name' #'tony','age' #30]

上述数据类型的值可以为 null。Pig 以与 SQL 类似的方式处理 null。null 可以是未知值或不存在值，也用作可选值的占位符。

Pig Latin 的结构运算符及其描述和示例如表 5-2 所示。

表 5-2 Pig Latin 的结构运算符及其描述和示例

结构运算符	描述	示例
()	元组构造函数运算符	(tony,30)
{}	包构造函数运算符	{(tony,30),(tom,45)}
[]	映射构造函数运算符	[name #tony,age #30]

Pig Latin 的关系运算符及其描述如表 5-3 所示。

表 5-3　Pig Latin 的关系运算符及其描述

关系运算符		描述
加载和存储	LOAD	将数据从文件系统（local/HDFS）加载到关系中
	STORE	将数据从文件系统（local/HDFS）存储到关系中
过滤	FILTER	从关系中删除不需要的行
	DISTINCT	从关系中删除重复行
	FOREACH、GENERATE	基于列数据生成数据转换
	STREAM	使用外部程序转换关系
分组和连接	JOIN	连接两个或多个关系
	COGROUP	将数据分组为两个或多个关系
	GROUP	在单个关系中对数据进行分组
	CROSS	创建两个或多个关系的向量积
排序	ORDER	基于一个或多个字段对数据按顺序（升序或降序）进行排序
	LIMIT	从关系中获取有限数量的元组
合并和拆分	UNION	将两个或多个关系合并为单个关系
	SPLIT	将单个关系拆分为两个或多个关系
诊断运算符	DUMP	在控制台中输出关系的内容
	DESCRIBE	描述关系的模式
	EXPLAIN	查看逻辑、物理计划或 MapReduce 执行计划以计算关系
	ILLUSTRATE	逐步执行语句序列

3. 输入、存储和输出

（1）输入

任何一种数据流的第一步都是指定输入。在 Pig Latin 中，使用 LOAD 语句来完成输入操作。默认情况下，如果不指定加载函数，则 LOAD 会使用默认加载函数 PigStorage() 加载存放在 HDFS 中且以制表符分隔的文件，如 divs=load 'pig_test'，其表示使用 PigStorage() 加载名为 pig_test 的文件。用户也可以通过指定一个完整的 URL 来加载文件，如 hdfs://xx.test.com/data/ pig_test，其表示可以从 NameNode 为 xx.test.com 的 HDFS 中读取文件。

实际上，用户的大部分数据并非是以制表符作为分隔符的文本文件，用户也有可能需要从其他非 HDFS 的存储系统中加载数据。Pig 允许用户在加载数据时通过 using 句式来指定其他加载函数。例如，从 HBase 中加载数据，语句如下。

```
divs = LOAD 'pig_test' USING HBasestorage();
```

如果没有指定加载函数，那么会使用内置的加载函数 PigStorage()。用户同样可以通过 using 句式为使用的加载函数指定分隔符。例如，如果想加载以逗号分隔的文本文件数据，那么 PigStorage()会接收一个指定分隔符的参数，语句如下。

```
divs = LOAD 'pig_test' USING PigStorage(',');
```

LOAD 语句还可以使用 as 句式作为用户加载的数据指定模式，这个模式相当于为数据集中的每个字段指定列名。

当从 HDFS 中访问指定文件的时候，用户也可以指定目录。在这种情况下，Pig 会遍历用户指定目录中的所有文件，并将它们作为 LOAD 语句的输入。

PigStorage()和 TextLoader()是内置的可操作 HDFS 文件的 Pig 加载函数，它们支持基于 Java 正则表达式的模式匹配语法，但并不完全支持所有正则表达式的特性。通过模式匹配，用户可以读取不在同一个目录中的多个文件，或者读取一个目录中的部分文件。

（2）存储

当用户处理完数据之后，需要把结果数据存储到某个地方。Pig 提供了使用 STORE 语句进行数据存储的方法。默认情况下，Pig 通过 PigStorage()将结果数据以制表符作为分隔符，并存储到 HDFS 中，此时需执行语句 "store processed into '/data';"。如果用户没有显式指定存储函数，那么将会默认使用 PigStorage()。用户可以使用 using 句式指定不同的存储函数，使用 using 句式指定 HBaseStorage()作为存储函数的语句如下。

```
STORE PROCESSED INTO '/data' using HBaseStorage();
```

用户也可以传递参数给其使用的存储函数。例如，如果想将数据存储为以逗号分隔的文本数据，则 PigStorage()会接收一个指定分隔符的参数，如执行语句 "store processed into '/data' using PigStorage(',');"。当数据存储到文件系统中后，data 目录中将包含多个文件，但是到底会生成多少个文件取决于执行 STORE 操作前的最后一个任务的并行数（该数由为这个任务所设置的并行级别所决定）。

（3）输出

用户可以使用 DUMP 语句将关于结果的数据输出到屏幕上，这条语句非常实用，特别是在调试阶段和原型研究阶段。可以 DUMP 语句将用户的脚本输出到屏幕上（即 "dump processed"）。

4. Pig Latin 常用操作

（1）查询指定行数据

```
tmp_table_limit = limit tmp_table 50;
dump tmp_table_limit;
```

（2）查询指定列数据

```
tmp_table_name = foreach tmp_table generate name;
dump tmp_table_name;
```

（3）为列取别名

```
tmp_table_column_alias = foreach tmp_table generate name as username, age as userage;
dump tmp_table_column_alias;
```

（4）按指定列进行排序

```
tmp_table_order = order tmp_table by age asc;
dump tmp_table_order;
```

（5）按条件进行查询

```
tmp_table_where = filter tmp_table by age > 18;
dump tmp_table_where;
```

（6）内连接

```
tmp_table_inner_join = join tmp_table by age, tmp_table2 by age;
dump tmp_table_inner_join;
```

（7）左连接

```
tmp_table_left_join = join tmp_table by age left outer, tmp_table2 by age;
dump tmp_table_left_join;
```

（8）右连接

```
tmp_table_right_join = join tmp_table by age right outer, tmp_table2 by age;
dump tmp_table_right_join;
```

（9）全连接

```
tmp_table_full_join = join tmp_table by age full outer, tmp_table2 by age;
dump tmp_table_full_join;
```

（10）交叉查询多张表

```
tmp_table_cross = cross tmp_table, tmp_table2;
dump tmp_table_cross;
```

（11）分组

```
tmp_table_group = group tmp_table by is_child;
dump tmp_table_group;
```

（12）分组并统计

```
tmp_table_group_count = group tmp_table by is_child;
tmp_table_group_count = foreach tmp_table_group_count generate group,count($1);
dump tmp_table_group_count;
```

（13）查询并去重

```
tmp_table_distinct = foreach tmp_table generate is_child;
tmp_table_distinct = distinct tmp_table_distinct;
dump tmp_table_distinct;
```

项目 5　Hadoop 生态组件的安装与使用

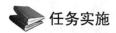

任务实施

1. Pig 的安装与配置

（1）安装所需的软件

微课 5-2　Pig 的安装与使用

① 在 master1 节点上切换到 Pig 安装包所在目录，执行 "ll" 命令查看目录，如图 5-4 所示。

```
[root@master1 opt]# ll
total 1788372
-rw-r--r--. 1 root root  55711670 Mar  2 18:15 apache-flume-1.7.0-bin.tar.gz
-rw-r--r--. 1 root root  92834839 Feb 17 11:27 apache-hive-1.2.1-bin.tar.gz
-rw-r--r--. 1 root root 218720521 Feb  6 00:03 hadoop-2.7.7.tar.gz
-rw-r--r--. 1 root root 121498244 Mar  2 08:48 hbase-1.6.0-bin.tar.gz
-rw-r--r--. 1 root root 232190985 Feb 19 09:11 hbase-2.2.2-bin.tar.gz
-rw-r--r--. 1 root root 148162542 Feb  6 00:03 jdk-8u341-linux-x64.tar.gz
-rw-r--r--. 1 root root 644399365 Feb 15 23:55 mysql-5.7.23-linux-glibc2.12-x86_64.tar.
drwxr-xr-x. 4 root root       151 Sep 25  2016 mysql-connector-java-5.1.40
-rw-r--r--. 1 root root   3911557 Feb 15 23:55 mysql-connector-java-5.1.40.tar.gz
-rw-r--r--. 1 root root 230606579 Mar  2 15:19 pig-0.17.0.tar.gz
-rw-r--r--. 1 root root  17953604 Feb 27 16:39 sqoop-1.4.7.bin__hadoop-2.6.0.tar.gz
-rw-r--r--. 1 root root      2100 Mar  3 16:09 text.txt
-rw-r--r--. 1 root root  22261552 Feb 15 23:56 zookeeper-3.4.8.tar.gz
```

图 5-4　查看目录

② 执行 "tar -zxvf pig-0.17.0.tar.gz -C /usr/local" 命令，解压 Pig 安装包到指定目录中，操作命令如下。

```
[root@master1 opt]# tar -zxvf pig-0.17.0.tar.gz -C /usr/local
```

③ 将解压文件重命名为 pig，操作命令如下。

```
[root@master1 local]# mv pig-0.17.0 pig
```

④ 若是普通用户，则可执行 "chown -R hadoop:hadoop pig" 命令，为 pig 目录修改权限，操作命令如下。

```
[root@master1 local]# chown -R hadoop:hadoop pig
```

（2）编辑环境变量

① 编辑 ~/.bashrc 文件，将 Pig 的安装路径添加到环境变量中，编辑内容如下。

```
export PIG_HOME=/usr/local/pig
PATH=$PATH:$PIG_HOME/bin
```

② 修改完成后，按 "Esc" 键返回命令行模式，通过输入 ":wq!" 保存修改并退出，随后执行 "source ~/.bashrc" 命令，使配置文件立即生效，操作命令如下。

```
[root@master1 local]# source ~/.bashrc
```

（3）启动测试

① 执行 "pig -x local" 命令进入本地模式，访问本地文件系统，测试或处理小规模数据集，如图 5-5 所示。

② 执行 "pig -x mapreduce" 命令进入 MapReduce 模式，在 MapReduce 模式下，Pig 可以访问整个 Hadoop 集群，处理大规模数据集，如图 5-6 所示。

```
[root@master1 local]# pig -x local
SLF4J: Class path contains multiple SLF4J bindings.
SLF4J: Found binding in [jar:file:/usr/local/hadoop/share/hadoop/common/lib/slf4j-l
SLF4J: Found binding in [jar:file:/usr/local/hbase/lib/slf4j-log4j12-1.7.25.jar!/or
SLF4J: See http://www.slf4j.org/codes.html#multiple_bindings for an explanation.
SLF4J: Actual binding is of type [org.slf4j.impl.Log4jLoggerFactory]
23/03/10 16:14:59 INFO pig.ExecTypeProvider: Trying ExecType : LOCAL
23/03/10 16:14:59 INFO pig.ExecTypeProvider: Picked LOCAL as the ExecType
2023-03-10 16:14:59,404 [main] INFO  org.apache.pig.Main - Apache Pig version 0.17.
2023-03-10 16:14:59,404 [main] INFO  org.apache.pig.Main - Logging error messages t
2023-03-10 16:14:59,418 [main] INFO  org.apache.pig.impl.util.Utils - Default bootu
2023-03-10 16:14:59,551 [main] INFO  org.apache.hadoop.conf.Configuration.deprecati
2023-03-10 16:14:59,552 [main] INFO  org.apache.pig.backend.hadoop.executionengine.
2023-03-10 16:14:59,653 [main] INFO  org.apache.hadoop.conf.Configuration.deprecati
2023-03-10 16:14:59,663 [main] INFO  org.apache.pig.PigServer - Pig Script ID for t
2023-03-10 16:14:59,663 [main] WARN  org.apache.pig.PigServer - ATS is disabled sin
grunt>
```

图 5-5 进入本地模式

```
[root@master1 local]# pig -x mapreduce
23/03/10 16:19:02 INFO pig.ExecTypeProvider: Trying ExecType : LOCAL
23/03/10 16:19:02 INFO pig.ExecTypeProvider: Trying ExecType : MAPREDUCE
23/03/10 16:19:02 INFO pig.ExecTypeProvider: Picked MAPREDUCE as the ExecType
2023-03-10 16:19:02,887 [main] INFO  org.apache.pig.Main - Apache Pig version 0.17.
0 (r1797386) compiled Jun 02 2017, 15:41:58
2023-03-10 16:19:02,887 [main] INFO  org.apache.pig.Main - Logging error messages t
o: /usr/local/pig_1678436342886.log
2023-03-10 16:19:02,901 [main] INFO  org.apache.pig.impl.util.Utils - Default bootu
p file /root/.pigbootup not found
SLF4J: Class path contains multiple SLF4J bindings.
SLF4J: Found binding in [jar:file:/usr/local/hadoop/share/hadoop/common/lib/slf4j-l
og4j12-1.7.10.jar!/org/slf4j/impl/StaticLoggerBinder.class]
SLF4J: Found binding in [jar:file:/usr/local/hbase/lib/slf4j-log4j12-1.7.25.jar!/or
g/slf4j/impl/StaticLoggerBinder.class]
SLF4J: See http://www.slf4j.org/codes.html#multiple_bindings for an explanation.
SLF4J: Actual binding is of type [org.slf4j.impl.Log4jLoggerFactory]
2023-03-10 16:19:03,200 [main] INFO  org.apache.hadoop.conf.Configuration.deprecati
on - mapred.job.tracker is deprecated. Instead, use mapreduce.jobtracker.address
2023-03-10 16:19:03,201 [main] INFO  org.apache.pig.backend.hadoop.executionengine.
HExecutionEngine - Connecting to hadoop file system at: hdfs://master1:9000
2023-03-10 16:19:03,521 [main] INFO  org.apache.pig.PigServer - Pig Script ID for t
he session: PIG-default-faedac4b-ee62-42f2-ab80-a558c1da46e2
2023-03-10 16:19:03,521 [main] WARN  org.apache.pig.PigServer - ATS is disabled sin
ce yarn.timeline-service.enabled set to false
grunt>
```

图 5-6 进入 MapReduce 模式

2. Pig 的应用

(1) 计算多维度组合下的平均值

假设有数据文件 data1.txt (各数值之间以制表符分隔)。现在要求计算在第 2、3、4 列的所有组合中,最后两列的平均值。执行 "cat data1.txt" 命令,查看 data1.txt 内容,操作命令及结果如下。

```
[root@master1 pig]# cat data1.txt
a 1 2 3 4.2 9.8
a 3 0 5 3.5 2.1
b 7 9 9 -   -
```

```
a 7 9 9 2.6 6.2
a 1 2 5 7.7 5.9
a 1 2 3 1.4 0.2
```

为了验证计算结果的正确性，在此先人工计算一下结果。首先，第 2、3、4 列有一个组合为(1,2,3)，即第一行和最后一行的数据。对于这个维度组合来说，最后两列的平均值分别为

$$（4.2+1.4）/2=2.8$$
$$（9.8+0.2）/2=5.0$$

可以发现，组合(7,9,9)有两行记录，即第 3、4 行，但是第 3 行数据的最后两列没有值，因此它不应该被用于平均值的计算，也就是说，在计算平均值时，第 3 行的数据是无效数据。所以(7,9,9)组合的最后两列的平均值为 2.6 和 6.2。

而对于第 2、3、4 列的其他维度组合来说，都分别只有一行数据，因此最后两列的平均值其实就是它们自身。

现在使用 Pig 来进行计算，并输出最终计算结果。

先进入本地调试模式（pig -x local），再依次输入 Pig 代码，操作命令及结果如下。

```
grunt>A = LOAD 'data1.txt' AS (col1:chararray, col2:int, col3:int, col4:int,
col5:double, col6:double);
grunt>B = GROUP A BY (col2, col3, col4);
grunt>C = FOREACH B GENERATE group, AVG(A.col5), AVG(A.col6);
grunt>DUMP C;

((1,2,3),2.8,5.0)
((1,2,5),7.7,5.9)
((3,0,5),3.5,2.1)
((7,9,9),2.6,6.2)
```

从最终计算结果可以看出，人工计算结果和 Pig 计算结果完全一致。代码分析如下。

① 加载 data1.txt 文件，并指定每一列的数据类型分别为 chararray、int、int、int、double、double。同时，分别给每一列指定一个别名，分别为 col1、col2、…、col6。别名在后面的数据处理中会用到，如果不指定别名，则在后面的数据处理中只能使用索引（$0,$1,…）来标示相应的列，这样命令可读性会变差。

将数据加载之后保存到变量 A 中，A 的数据结构如下。

```
A: {col1: chararray,col2: int,col3: int,col4: int,col5: double,col6: double}
```

可见 A 是用"{}"标识的包。

② 按照 A 的第 2、3、4 列对 A 进行分组。Pig 会找出第 2、3、4 列的所有组合，并按照升序进行排列，将它们与对应的包 A 整合起来，得到如下数据结构。

```
B: {group: (col2:int,col3:int,col4:int),A: {col1:chararray,col2:int,col3:int,
col4:int,col5: double,col6:double}}
```

可见，A 的第 2、3、4 列的组合被 Pig 赋予了一个别名（group），这很形象。同时，可以观察到，B 的每一行其实就是由一个 group 和若干个 A 组成的。B 的实际数据如下。

```
((1,2,3),{(a,1,2,3,4.2,9.8),(a,1,2,3,1.4,0.2)})
((1,2,5),{(a,1,2,5,7.7,5.9)})
((3,0,5),{(a,3,0,5,3.5,2.1)})
((7,9,9),{(b,7,9,9,,),(a,7,9,9,2.6,6.2)})
```

可见,组合(1,2,3)对应了两行数据,组合(7,9,9)也对应了两行数据。

③ 计算每一种组合下的最后两列的平均值。

根据得到的 B 的数据,可以把 B 想象为一行一行的数据(这些行不是对称的),FOREACH 的作用是对 B 的每一行数据进行遍历并计算。GENERATE 表示指定要生成什么样的数据,这里的 group(即 Pig 为 A 的第 2、3、4 列的组合赋予的别名)就是第②步操作中 B 的第一项数据,所以在数据集 C 的每一行中,第一项就是 B 中的 group,类似于组合(1,2,3)的形式。而 AVG(A.col5)调用了 Pig 的求平均值的函数 AVG(),用于对 A 的名为 col5 的列(别名,col5 就是倒数第二列)求平均值。

在此操作中遍历的是 B,在 B 的数据结构中,每一行数据中都包含一个 group,其对应的是若干个 A,因此这里的 A.col5 中的 A 指的是 B 的每一行中的 A,而不是包含全部数据的 A。例如,((1,2,3),{(a,1,2,3,4.2,9.8),(a,1,2,3,1.4,0.2)}),遍历到 B 的这一行时,要计算 AVG(A.col5),Pig 会找到(a,1,2,3,4.2,9.8) 中的 4.2 及(a,1,2,3,1.4,0.2)中的 1.4,将其加起来除以 2,这样就得到了平均值。

同理,可以清楚地知道 AVG(A.col6)是怎样计算出来的。但有一点需要注意,对于(7,9,9)组合,它对应的数据(b,7,9,9,,)中的最后两列是没有值的,这是因为数据文件对应位置上不是有效数字,而是两个"-",Pig 在加载数据的时候自动将其置为空,且在计算平均值的时候,不会把这一组数据考虑在内(相当于忽略这组数据的存在)。

C 的数据结构如下。

```
C: {group: (col2: int,col3: int,col4: int),double,double}
```

④ DUMP C 表示将 C 中的数据输出到控制台上。如果要将数据输出到文件中,则需要使用以下语句。

```
STORE C INTO 'output';
```

这样 Pig 就会在当前目录中新建一个 output 目录(该目录必须事先不存在),并把结果文件放到该目录中。

(2)统计数据行数

在 SQL 语句中,统计表中数据的行数非常简单,使用以下语句即可。

```
SELECT COUNT(*) FROM table_name WHERE condition
```

Pig 中也有一个 COUNT()函数,假设要计算数据文件 data1.txt 的行数,可否使用以下语句进行操作呢?

```
A = LOAD 'data1.txt' USING PigStorage(' ')AS (col1:chararray, col2:int, col3:int,
col4:int, col5:double, col6:double);
B = COUNT(*);
DUMP B;
```

可以发现,上述操作会报错,且 Pig 手册中有如下信息。

```
Note: You cannot use the tuple designator (*) with COUNT; that is, COUNT(*) will
not work.
```

项目 5　Hadoop 生态组件的安装与使用

而修改语句为"B = COUNT(A.col2);"后，操作依然会报错。

要想统计 A 中含 col2 字段的数据有多少行，正确的做法是使用以下语句。

```
A = LOAD 'data1.txt' USING PigStorage(' ')AS (col1:chararray, col2:int, col3:int,
col4:int, col5:double, col6:double);
B = GROUP A ALL;
C = FOREACH B GENERATE COUNT(A.col2);
DUMP C;
(6)
```

在这个例子中，COUNT(A.col2)和 COUNT(A)的结果是一样的，但是当 col2 列中含有空值时，如这里假设 test.txt 的数据如下。

```
[root@localhost pig]$ cat test.txt
a 1 2 3 4.2 9.8
a   0 5 3.5 2.1
b 7 9 9 - -
a 7 9 9 2.6 6.2
a 1 2 5 7.7 5.9
a 1 2 3 1.4 0.2
```

Pig 程序的执行结果如下。

```
A = LOAD 'test.txt' USING PigStorage(' ')AS (col1:chararray, col2:int, col3:int,
col4:int, col5:double, col6:double);
B = GROUP A ALL;
C = FOREACH B GENERATE COUNT(A.col2);
DUMP C;
(5)
```

可见，结果为 5 行。这是因为加载数据的时候指定了 col2 的数据类型为 int，而 test.txt 中，第 2 行 col2 列的数据是空的，因此数据加载到 A 中后，有一个字段是空的。

```
grunt> DUMP A;
(a,1,2,3,4.2,9.8)
(a,,0,5,3.5,2.1)
(b,7,9,9,,)
(a,7,9,9,2.6,6.2)
(a,1,2,5,7.7,5.9)
(a,1,2,3,1.4,0.2)
```

在使用 COUNT()函数统计行数的时候，空字段不会被计入在内，所以结果是 5。

（3）对元组和包进行"解嵌套"

仍然使用前面的 data1.txt 数据文件来说明 FLATTEN 运算符，如果计算多维度组合下的最后两列的平均值，则语句如下。

```
A = LOAD 'data1.txt' USING PigStorage(' ')
    AS(col1:chararray,col2:int,col3:int,col4:int,col5:double,col6:double);
B = GROUP A BY (col2,col3,col4);
C = FOREACH B GENERATE group, AVG(A.col5), AVG(A.col6);
```

```
DUMP C;
((1,2,3),2.8,5.0)
((1,2,5),7.7,5.9)
((3,0,5),3.5,2.1)
((7,9,9),2.6,6.2)
```

可见,在输出结果中,每一行的第一项是一个元组。下面使用以下语句查看 FLATTEN 运算符的作用。

```
A = LOAD 'data1.txt' USING PigStorage(' ')
    AS(col1:chararray,col2:int,col3:int,col4:int,col5:double,col6:double);
B = GROUP A BY (col2, col3, col4);
C = FOREACH B GENERATE FLATTEN(group), AVG(A.col5), AVG(A.col6);
DUMP C;
(1,2,3,2.8,5.0)
(1,2,5,7.7,5.9)
(3,0,5,3.5,2.1)
(7,9,9,2.6,6.2)
```

结果显示,使用了 FLATTEN 运算符的 group 由元组变为扁平结构。按照 Pig 文档的说法,FLATTEN 运算符用于对元组和包进行"解嵌套"。

```
The FLATTEN operator looks like a UDF syntactically, but it is actually an operator
that changes the structure of tuples and bags in a way that a UDF cannot. Flatten
un-nests tuples as well as bags. The idea is the same, but the operation and result
is different for each type of structure.
For tuples, flatten substitutes the fields of a tuple in place of the tuple. For
example, consider a relation that has a tuple of the form (a, (b, c)). The expression
GENERATE $0, flatten($1), will cause that tuple to become (a, b, c).
```

有时,不"解嵌套"的数据是不利于观察的,如可能不利于外围程序的处理(例如,Pig 将数据输出到磁盘中后,如果需要使用其他程序做后续处理,则对于一个元组而言,其输出的内容中是含括号的,这就在处理流程上需要增加一道去除括号的工序),因此,FLATTEN 运算符为用户提供了一个在某些情况下可以清楚、方便地分析数据的机会。

(4)把数据当作"元组"来加载

此处依然使用 data1.txt 的数据来进行相关语句操作,首先,按照以下方式来加载数据。

```
A = LOAD 'data1.txt' USING PigStorage(' ')
    AS (col1:chararray,col2:int,col3:int,col4:int,col5:double,col6:double);
```

可得到 A 的数据结构如下。

```
DESCRIBE A;
A:{col1:chararray,col2:int,col3:int,col4:int,col5:double,col6:double}
```

如果想要把 A 当作元组来加载,则需要使用如下语句。

```
A = LOAD 'data1.txt' USING PigStorage(' ')
    AS (T:tuple(col1:chararray,col2:int,col3:int,col4:int,col5:double,
col6:double));
```

即想要得到 A 的数据结构如下。
```
DESCRIBE A;
A: {T: (col1: chararray,col2:int,col3:int,col4:int,col5:double,col6:double)}
```
但是，使用以上方法将得到一个空的 A。
```
grunt> DUMP A;
()
()
()
()
()
()
```
这是因为数据文件 data1.txt 的结构不适合将其作为元组来加载。如果数据文件 data2.txt 的数据如下。
```
[root@localhost pig]$ cat data2.txt
(a,1,2,3,4.2,9.8)
(a,3,0,5,3.5,2.1)
(b,7,9,9,-,-)
(a,7,9,9,2.6,6.2)
(a,1,2,5,7.7,5.9)
(a,1,2,3,1.4,0.2)
```
则使用上面的加载方法进行加载后，A 的数据结构及内容。
```
A = LOAD 'data2.txt' AS (T:tuple (col1:chararray,col2:int,col3:int,col4:
int,col5:double, col6:double));
DUMP A;
((a,1,2,3,4.2,9.8))
((a,3,0,5,3.5,2.1))
((b,7,9,9,,))
((a,7,9,9,2.6,6.2))
((a,1,2,5,7.7,5.9))
((a,1,2,3,1.4,0.2))
```
可见，加载的数据的结构确实被定义为了元组。

（5）在多维度组合下，计算某个维度组合中不重复记录的条数

以数据文件 data3.txt 为例，计算在第 2、3、4 列的所有维度组合下，最后一列不重复的记录条数，data3.txt 的内容如下。
```
[root@localhost pig]$ cat data3.txt
a 1 2 3 4.2 9.8 100
a 3 0 5 3.5 2.1 200
b 7 9 9 - - 300
a 7 9 9 2.6 6.2 300
a 1 2 5 7.7 5.9 200
a 1 2 3 1.4 0.2 500
```

可以发现，第 2、3、4 列有一个维度组合是(1,2,3)，在这个维度组合下，最后一列有两种值——100 和 500，因此不重复记录的条数为 2。同理，可求得其他不重复记录的条数。Pig 程序的执行结果如下。

```
A = LOAD 'data3.txt' USING PigStorage(' ')AS (col1:chararray, col2:int, col3:int,
col4:int, col5:double, col6:double, col7:int);
B = GROUP A BY (col2, col3, col4);
C = FOREACH B {D = DISTINCT A.col7; GENERATE group, COUNT(D);};
DUMP C;
((1,2,3),2)
((1,2,5),1)
((3,0,5),1)
((7,9,9),1)
```

代码分析如下。

① LOAD 表示加载数据。

② GROUP 的作用和前面表述的一样。完成 GROUP 操作之后得到的 B 的数据如下。

```
grunt> DUMP B;
((1,2,3),{(a,1,2,3,4.2,9.8,100),(a,1,2,3,1.4,0.2,500)})
((1,2,5),{(a,1,2,5,7.7,5.9,200)})
((3,0,5),{(a,3,0,5,3.5,2.1,200)})
((7,9,9),{(b,7,9,9,,,300),(a,7,9,9,2.6,6.2,300)})
```

③ DISTINCT 用于将一个关系中重复的元组移除，FOREACH 用于对 B 的每一行进行遍历，其中 B 的每一行中都含有一个包，每一个包中都含有若干元组 A。因此，FOREACH 后面"{}"中的操作其实是对所谓的"内部包"的操作，这里指定了对 A 的 col7 列进行去重，去重的结果被命名为 D，对 D 进行计数（COUNT()），最终得到了想要的结果。

④ DUMP 表示使结果数据进行输出显示。

（6）使用 Shell 进行辅助数据处理

Pig 中可以嵌套使用 Shell 进行辅助数据处理，下面以 data4.txt 为例对该方法进行介绍，data4.txt 中的数据如下。

```
[root@localhost pig]$ cat data4.txt
1 5 98 = 7
34 8 6 3 2
62 0 6 = 65
```

如果想要将 data4.txt 内第 4 列中的"="全部替换为 9999，可以使用如下 Pig 程序来完成。

```
A = LOAD 'data4.txt' USING PigStorage(' ') AS (col1:int,col2:int,col3:int,
col4:chararray,col5:int);
B = STREAM A THROUGH `awk '{if($4 == "=") print $1" "$2" "$3" 9999 "$5; else print
$0;}'`;
DUMP B;
(1,5,98,9999,7)
```

项目 5 Hadoop 生态组件的安装与使用

```
(34,8,6,3,2)
(62,0,6,9999,65)
```

代码分析如下。

① LOAD 表示加载数据。

② 通过 "STREAM … THROUGH …" 的方式，可以调用一个 Shell 语句，使用该 Shell 语句能够对 A 的每一行数据进行处理。此处的 Shell 逻辑是，若某一行数据的第 4 列为 "="，将其替换为 "9999" 并输出；否则按照原样输出。

③ DUMP 表示输出 B。

（7）向 Pig 脚本中传入参数

若 Pig 脚本输出的文件是通过外部参数指定的，则此参数需要传入。在 Pig 中，传入参数的语句如下。

```
STORE A INTO '$output_dir';
```

其中，"output_dir" 就是传入的参数。在调用 Pig 的 Shell 脚本时，可以使用以下语句传入参数。

```
pig -param output_dir="/home/my_output_dir/" my_pig_script.pig
```

这里传入的参数 "output_dir" 的值为 "/home/my_output_dir/"。

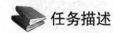 Flume 的安装与使用

任务描述

学习 Flume 的相关知识、熟悉 Flume 的作用、完成 Flume 的安装与配置等。

任务目标

（1）学会 Flume 的安装与配置。
（2）学会使用 Flume 将日志数据上传至 HDFS。

知识准备

Flume 作为 Cloudera 开发的实时日志收集系统，受到了业界的认可与广泛应用。Flume 初始的发行版本目前被统称为 Flume OG（Original Generation），属于 Cloudera。

随着 Flume 功能的扩展，Flume OG 代码工程臃肿、核心组件设计不合理、核心配置不标准等缺点逐渐暴露出来，尤其是在 Flume OG 的最后一个发行版本 0.9.4 中，日志传输不稳定的现象尤为严重。为了解决这些问题，Cloudera 完成了 Flume-728 的开发，对 Flume 进行了里程碑式的改动，重构了其核心组件、核心配置及代码架构，重构后的版本被统称为 Flume NG（Next Generation）；这一改动的另一原因是将 Flume 纳入 Apache 旗下，Cloudera Flume 改名为 Apache Flume。

Flume 是一个分布式的、可靠的、高可用的海量日志收集、聚合和传输系统。其支持在日志系统中定制各类数据发送方，用于收集数据。同时，Flume 提供对数据进行简单处理，

并写到各类数据接收方（如文本、HDFS、HBase 等）的功能。

Flume 的数据流由事件（Event）贯穿始终。事件是 Flume 的基本数据单位，它包含日志数据（字节数组形式）及头信息，这些事件由 Agent 外部的 Source 生成，当 Source 捕获事件后，会对其进行特定的格式化，且 Source 会把事件推入单个或多个 Channel 中。可以把 Channel 看作一个缓冲区，它将保存事件直到 Sink 处理完该事件。Sink 负责持久化日志或者把事件推向另一个 Source。

Flume 主要由以下 3 个重要的组件构成。

① Source。完成对日志数据的收集，负责捕获数据并进行特殊格式化，将数据封装到事件里然后将事件推入 Channel 中。Flume 提供了 Source 的各种实现，包括 Avro Source、Exce Source、SpoolingDirectory Source、NetCat Source、Syslog Source、Syslog TCP Source、Syslog UDP Source、HTTP Source、HDFS Source 等。

② Channel。主要提供队列的功能，对 Source 提供的数据进行简单的缓存。Flume 中的 Channel 包括 Memory Channel、JDBC Channel、File Channel 等。

③ Sink。用于取出 Channel 中的数据，并将其存储到文件系统、数据库或者提交到远程服务器中。Flume 中的 Sink 包括 HDFS Sink、Logger Sink、Avro Sink、File Roll Sink、Null Sink、HBase Sink 等。

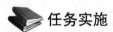 任务实施

1. Flume 的安装与配置

（1）安装所需要的软件

① 在 Flume 安装包目录中，执行 "tar -zxvf /opt/apache-flume-1.7.0-bin.tar.gz -C /usr/local/" 命令，解压 Flume 安装包到指定目录中，操作命令如下。

微课5-3 Flume 的安装与使用

```
[root@master1 opt]# tar -zxvf /opt/apache-flume-1.7.0-bin.tar.gz -C /usr/local/
```

② 执行 "mv apache-flume-1.7.0-bin flume" 命令，将解压文件重命名为 flume，操作命令如下。

```
[root@master1 local]# mv apache-flume-1.7.0-bin flume
```

若是普通用户，则可通过执行 "chown -R hadoop:hadoop /usr/local/flume/" 命令修改 flume 目录的权限，操作命令如下。

```
[root@master1 conf]# chown -R hadoop:hadoop /usr/local/flume/
```

（2）编辑环境变量

① 编辑 ~/.bashrc 文件，将 HBase 的安装路径添加到环境变量中，编辑内容如下。

```
export FLUME_HOME=/usr/local/flume
PATH=$PATH:$SQOOP_HOME/bin:$PIG_HOME/bin:$FLUME_HOME/bin
```

② 修改完成后，按 "Esc" 键返回命令行模式，通过输入 ":wq!" 保存修改并退出，随后执行 "source ~/.bashrc" 命令，使配置文件立即生效，操作命令如下。

```
[root@master1 local]# source ~/.bashrc
```

③ 将配置文件 flume-env.sh.template 重命名为 flume-env.sh，并编辑 flume-env.sh 配置文件，修改配置项内容如下。

项目 5　Hadoop 生态组件的安装与使用

```
# Enviroment variables can be set here.
 export JAVA_HOME=/usr/lib/jdk1.8.0
```

④ 配置完成后，进入 flume/bin 目录，执行 "./flume-ng version" 命令，查看 Flume 是否安装成功，操作命令及结果如下。

```
[root@master1 bin]# ./flume-ng version
Error: Could not find or load main class org.apache.flume.tools.GetJavaProperty
Flume 1.7.0
Source code repository: https://git-wip-us.apache.org/repos/asf/flume.git
Revision: 511d868555dd4d16e6ce4fedc72c2d1454546707
Compiled by bessbd on Wed Oct 12 20:51:10 CEST 2016
From source with checksum 0d21b3ffdc55a07e1d08875872c00523
```

⑤ 启动 Flume 后会提示 "org.apache.flume.tools.GetJavaProperty" 错误，可以通过修改 hbase-env.sh 文件来解决这个错误。执行 "vi hbase-env.sh" 命令，使用编辑器打开 hbase-env.sh 文件，操作命令如下。

```
[root@master1 conf]# vi hbase-env.sh
```

⑥ 编辑 hbase-env.sh 文件，在 hbase-env.sh 文件中的 "HBASE_CLASSPATH=/usr/local/hbase/conf" 一行的最前端加上注释符号。

⑦ 重新启动 Flume，操作命令及结果如下。

```
[root@master1 bin]# flume-ng version
Flume 1.7.0
Source code repository: https://git-wip-us.apache.org/repos/asf/flume.git
Revision: 511d868555dd4d16e6ce4fedc72c2d1454546707
Compiled by bessbd on Wed Oct 12 20:51:10 CEST 2016
From source with checksum 0d21b3ffdc55a07e1d08875872c00523
```

2. 使用 Flume 将日志数据上传至 HDFS

① 执行 "start-all.sh" 命令，启动 Hadoop 集群，在选定目录中创建一个文件，如在 /simple 目录中执行 "touch a2.conf" 命令，创建 a2.conf 文件。需要注意，创建 a2.conf 文件的目录需要与配置文件中的 "a2.sources.r1.command" 配置项的值一致，编辑配置文件 a2.conf 如下。

```
a2.sources = r1
a2.channels = c1
a2.sinks = k1

a2.sources.r1.type = exec
a2.sources.r1.command = tail -F /simple/data.txt

a2.channels.c1.type = memory
a2.channels.c1.capacity = 1000
a2.channels.c1.transactionCapacity = 100
```

```
a2.sinks.k1.type = hdfs
a2.sinks.k1.hdfs.path = hdfs://172.18.0.2:9000/flume/date_hdfs.txt
a2.sinks.k1.hdfs.filePrefix = events-
a2.sinks.k1.hdfs.fileType = DataStream

a2.sources.r1.channels = c1
a2.sinks.k1.channel = c1
```

② 进入 flume/bin 目录中，使用配置文件 a2.conf 启动 Flume，操作命令如下。

```
[root@master1 bin]# ./flume-ng agent -n a2 -f /simple/a2.conf -c ../conf/ -Dflume.root.logger=INFO,console
```

③ 创建任意测试数据文件 data2.txt，模拟日志的生成，操作命令如下。

```
[root@master1 simple]# cat data2.txt >> data.txt
```

④ 执行 "hadoop fs -cat /flume/date_hdfs.txt/*" 命令，查看 HDFS 中生成的文件内容，操作命令及结果如下。

```
[root@master1 simple]# hadoop fs -cat /flume/date_hdfs.txt/*
(a,1,2,3,4.2,9.8)
(a,3,0,5,3.5,2.1)
(b,7,9,9,-,-)
(a,7,9,9,2.6,6.2)
(a,1,2,5,7.7,5.9)
(a,1,2,3,1.4,0.2)
(a,1,2,3,4.2,9.8)
(a,3,0,5,3.5,2.1)
(b,7,9,9,-,-)
(a,7,9,9,2.6,6.2)
(a,1,2,5,7.7,5.9)
(a,1,2,3,1.4,0.2)
```

任务 5.4　Kafka 的安装与使用

任务描述

学习 Kafka 的相关知识、熟悉 Kafka 的作用、完成 Kafka 的安装与配置等。

任务目标

（1）学会 Kafka 的安装与配置。
（2）理解 Kafka 的工作原理。

知识准备

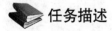

Kafka 是一个支持分区、多副本，且由 ZooKeeper 进行协调的分布式消息系统。它由服

项目 5　Hadoop 生态组件的安装与使用

务器和客户端组成，通过高性能传输控制协议（Transmission Control Protocol，TCP）进行通信。它可以部署在本地和云环境中的裸机硬件、虚拟机和容器上。

Kafka 的服务器和客户端的主要作用如下。

服务器：Kafka 作为由一个或多个服务器的集群来运行，这些服务器可以跨越多个数据中心或云区域。其中一些服务器形成了存储层，被称为代理；其他服务器运行 Kafka Connect，将数据作为事件流不断导入和导出，以将 Kafka 与现有系统（如关系数据库和其他 Kafka 集群）集成。

客户端：允许用户编写分布式应用程序和微服务，即使在出现网络问题或机器故障的情况下，客户端也可以并行地、大规模地、有容错地读取、写入和处理事件流。

Kafka 是一个事件流平台，其以事件流的形式从数据库、传感器、移动设备、云服务和软件应用程序等事件源中实时捕获数据，持久地存储这些事件流以供以后检索，实时地以及回顾性地操纵、处理和响应事件流，或根据需要将事件流路由到不同的目的地，Kafka 最大的特性就是可以实时处理大量数据以满足多种需求场景。Kafka 可以与 Flume、Storm、Spark Streaming、HBase、Flink 和 Spark 协同工作，对流数据进行实时读取、分析和处理，其应用实例如下。

（1）实时处理支付等金融交易，如应用于证券、银行和保险行业。

（2）实时跟踪和监控汽车、卡车、车队及货运信息，如应用于物流和汽车行业。

（3）持续捕获和分析来自物联网设备或其他设备（如工厂和风电场的设备）的传感器数据。

（4）收集用户互动信息和订单信息并立即做出反应，如应用于零售、酒店和旅游业。

（5）监测医院护理中的患者并预测病情变化，以确保在紧急情况下及时对其进行治疗。

（6）连接、存储并提供由企业不同部门生成的数据。

（7）作为数据平台、事件驱动架构和微服务的基础。

一个事件记录了"发生的一些事情"的事实，也称为记录或消息。读或写数据到 Kafka 的过程是以事件的形式来完成的。从概念上讲，事件由键（Key）、值（Value）、时间戳（Timestamp）和可选的元数据（Metadata）头组成。可选的元数据头类似于 HTTP 或者 TCP/IP 中的 Header 头部，可在消息中添加一些描述性信息，方便消费者解析和处理消息。事件示例如下。

事件的键："佩奇"。

事件的值："向乔治支付了 100 元"。

事件的时间戳："2023 年 1 月 1 日下午 1：11"。

Header 的 Key：消息 ID。

Header 的 Value：20231122。

生产者（Producer）是那些向 Kafka 发布（写入）事件的客户端应用程序，消费者（Consumer）是那些订阅（读取和处理）这些事件的人、应用程序等。在 Kafka 中，生产者和消费者是完全解耦的，彼此不可知，这是实现 Kafka 高可扩展性的关键。例如，生产者永远不需要等待消费者。

事件按主题（Topic）进行组织和持久存储。主题类似于文件系统中的目录，事件是该目录中的文件。例如，主题名称可以是"支付"。Kafka 中的主题总是多生产者和多消费者，即一个主题可以有 0 个、1 个或多个向其写入事件的生产者,也可以有 0 个、1 个或多个订阅这些事件的消费者,消费者可以根据需要随时读取主题中的事件。与传统的消息传递系统不同,Kafka 在使用后不会删除事件。用户可以通过对每个主题的配置进行设置来定义 Kafka 保留事件的时间,超过该时间后旧事件将被丢弃。

主题是分区的,这意味着一个主题分布在位于不同 Kafka 代理上的多个"桶"上。这种数据的分布式放置对于可扩展性非常重要,因为它允许客户端应用程序同时从多个代理读取数据和向多个代理写入数据。如图 5-7 所示,当一个新事件发布到一个主题时,它实际上被写入该主题的一个分区。具有相同键(用相同颜色表示)的事件被写入同一分区,Kafka 保证给定主题分区的任何消费者将始终以与写入时完全相同的顺序读取该分区的事件。

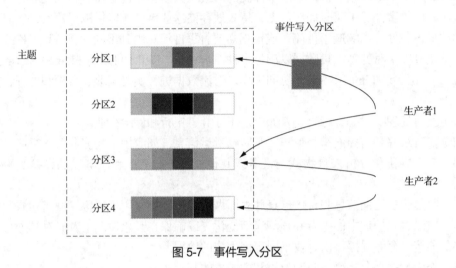

图 5-7 事件写入分区

为了使数据具有容错性和高可用性,Kafka 的每个主题都可以被复制,复制时甚至可以跨地理区域或数据中心,这样可保证总是有多个代理拥有数据副本,默认数据副本参数为 3,即始终有 3 份数据副本,该复制是在主题分区级别执行的。

除了用于操作和管理任务的命令行工具外,Kafka 还有 5 个用于 Java 和 Scala 的核心 API。

① Admin API。用于管理和查看主题、代理与其他 Kafka 对象。

② Producer API。用于向一个或多个 Kafka 主题发布(写入)事件流。

③ Consumer API。用于订阅(读取)一个或多个主题并处理为其生成的事件流。

④ Kafka Streams API。用于实现流处理应用程序和微服务。它提供了更高级别的事件流处理相关的功能,包括转换、聚合、连接等。它通过从一个或多个主题读取输入,生成一个或更多主题的输出,从而有效地将输入流转换为输出流。

⑤ Kafka Connect API。用于构建和运行可重复使用的数据导入/导出连接器,这些连接器能够从外部系统和应用程序上消费(读取)或生产(写入)事件流,以便与 Kafka 集成。在实践中,用户通常不需要实现自己的连接器,因为 Kafka 社区已经提供了数百个现成的连接器。

项目 5 Hadoop 生态组件的安装与使用

任务实施

1. 安装与配置 Kafka

(1) 运行 Docker 容器,并分别启用 master1、slave1 和 slave2 节点的 ZooKeeper 服务。

微课 5-4 Kafka 的安装与使用

在 master1 节点上执行 "jps" 命令查看进程,操作命令及结果如下。

```
[root@master1 bin]# jps
135 QuorumPeerMain
169 Jps
```

在 slave1 节点上执行 "jps" 命令查看进程,操作命令及结果如下。

```
[root@slave1 bin]# jps
145 Jps
115 QuorumPeerMain
```

在 slave2 节点上执行 "jps" 命令查看进程,操作命令及结果如下。

```
[root@slave2 bin]# jps
104 QuorumPeerMain
142 Jps
```

(2) 解压 Kafka 安装包,切换到解压目录并将文件重命名为 kafka,操作命令如下。

```
[root@master1 opt]# tar -zxvf /opt/kafka_2.11-0.10.2.0.tgz -C /usr/local/
[root@master1 local]# mv kafka_2.11-0.10.2.0/ kafka
```

(3) 编辑 ~/.bashrc 文件,将 Kafka 的安装路径添加到环境变量中,编辑内容如下。

```
export KAFKA_HOME=/usr/local/kafka
PATH=$PATH:$KAFKA_HOME/bin
```

修改完成后,按 "Esc" 键返回命令行模式,通过输入 ":wq!" 保存修改并退出,随后执行 "source ~/.bashrc" 命令,使配置文件立即生效,操作命令如下。

```
[root@master1 local]# source ~/.bashrc
```

(4) 进入 kafka/config 目录,修改 server.properties 文件,操作命令和修改配置项内容如下。

```
[root@master1 config]# vi server.properties
#修改如下两个配置项
log.dirs=/usr/local/kafka/logs
zookeeper.connect=master1:2181,slave1:2181,slave2:2181
#在文件最后增加如下两行内容
host.name=master1
delete.topic.enable=true
```

其中,/usr/local/kafka/logs 目录需要手动创建,切换到 Kafka 的安装目录中,创建 logs 目录,操作命令如下。

```
[root@master1 kafka]# mkdir logs
```

(5) 将配置好的 Kafka 目录分发给集群中的 slave1 和 slave2 节点,操作命令如下。

```
[root@master1 kafka]# scp -r /usr/local/kafka slave1:/usr/local/
[root@master1 kafka]# scp -r /usr/local/kafka slave2:/usr/local/
```
在 slave1 节点上修改 server.properties 文件,修改内容如下。
```
broker.id=1
host.name=slave1
```
在 slave2 节点上修改 server.properties 文件,修改内容如下。
```
broker.id=2
host.name=slave2
```
(6)在 master1、slave1 和 slave2 节点上,分别切换到 kafka/bin 目录,启动 Kafka(需要在 ZooKeeper 已启用的前提下),并分别执行 "jps" 命令查看进程。

在 master1 节点上的操作命令及结果如下。
```
[root@master1 bin]# ./kafka-server-start.sh -daemon ../config/server.properties
[root@master1 bin]# jps
1632 Kafka
135 QuorumPeerMain
1694 Jps
```
在 slave1 节点上的操作命令及结果如下。
```
[root@slave1 bin]# ./kafka-server-start.sh -daemon ../config/server.properties
[root@slave1 bin]# jps
115 QuorumPeerMain
419 Kafka
477 Jps
```
在 slave2 节点上的操作命令及结果如下。
```
[root@slave2 bin]# ./kafka-server-start.sh -daemon ../config/server.properties
[root@slave2 bin]# jps
104 QuorumPeerMain
490 Jps
431 Kafka
```

2. Kafka 运行测试

(1)在 master1 节点上创建主题,并在集群中查看已创建的主题,操作命令及结果如下。
```
[root@master1 bin]# ./kafka-topics.sh --create --zookeeper master1:2181,
slave1:2181,slave2:2181 --replication-factor 3 --partitions 3 --topic test
Created topic "test".

[root@master1 bin]# ./kafka-topics.sh --list --zookeeper master1:2181, slave1:
2181, slave2:2181
test
```
(2)在 master1 节点上启动生产者并测试,操作命令及结果如下。
```
[root@master1 bin]# ./kafka-console-producer.sh --broker-list master1:9092,
slave1:9092,slave2:9092 --topic test
>book
```

项目 5　Hadoop 生态组件的安装与使用

```
>test
>
```

（3）在 slave1 或 slave2 节点上启动消费者，消费者会自动输出步骤（2）中 test 主题输入的内容，操作命令及结果如下。

```
[root@slave1 bin]./kafka-console-consumer.sh --bootstrap-server master1:9092,
slave1:9092,slave2:9092 --from-beginning --topic test
[root@slave1 bin]# ./kafka-console-consumer.sh --bootstrap-server master1:9092,
slave1:9092,slave2:9092 --from-beginning --topic test
book
test
```

任务 5.5　Flink 的安装与使用

任务描述

学习 Flink 的相关知识、熟悉 Flink 的作用、完成 Flink 的安装与配置等。

任务目标

（1）学会 Flink 的安装与配置。
（2）理解 Flink 主要进程的作用，学会 Flink 的操作。

知识准备

Apache Flink 是一个面向数据流处理和批量数据处理的分布式开源计算框架，它基于 Flink 流式执行模型（Streaming Execution Model），能够支持流处理和批处理两种应用类型，适用于各种实时数据处理场景，可以对有界数据集和无界数据流进行高效、准确、可靠的实时计算及数据处理。其中，有界数据集指的是有限大小的数据集；而无界数据流指的是无限的数据流，如来自传感器、日志、消息队列等的数据。

Flink 框架如图 5-8 所示。

Flink 的部署（Deploy）支持单机模式（Local）和集群模式（Standalone、HA、YARN），也支持云端部署（GCE、EC2）。

Flink 的核心（Core）是分布式数据流引擎（Runtime）。

Flink 的 API 有 DataStream API 和 DataSet API，分别用于流处理和批处理。DataStream API 支持 CEP、Table 操作和 SQL 操作。DataSet API 支持机器学习库（FlinkML）、图计算（Gelly）、Table 操作和 SQL 操作。

Flink 包含三大核心组件，如图 5-9 所示。

① 数据源（Data Source）。负责接收数据。
② 算子（Transformation）。负责对数据进行处理。
③ 输出组件（Data Sink）。负责把计算好的数据输出到其他存储介质中。

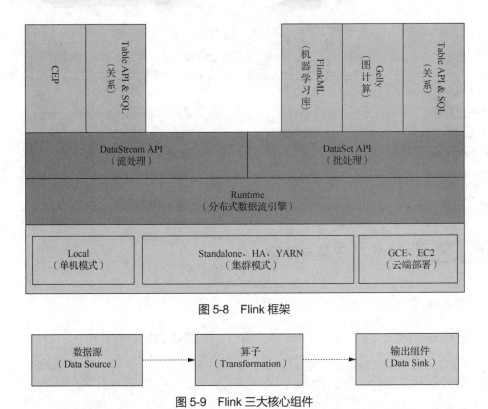

图 5-8　Flink 框架

图 5-9　Flink 三大核心组件

Flink 系统架构如图 5-10 所示。

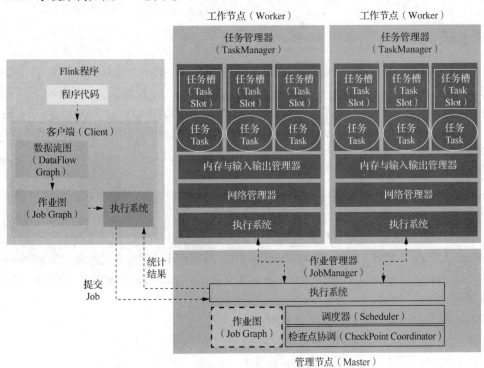

图 5-10　Flink 系统架构

项目 5　Hadoop 生态组件的安装与使用

Flink 集群启动时，会启动一个作业管理器（JobManager）进程和至少一个任务管理器（TaskManager）进程。在单机模式下，会在同一个 Java 虚拟机内部启动一个 JobManager 进程和 TaskManager 进程。JobManager 负责接收并管理作业（Job），TaskManager 负责执行任务（Task）。

一个作业由一个或多个任务组成，这些任务可以分配到不同的 TaskManager 上运行。TaskManager 包含多个任务槽（Task Slot），每个任务槽都可以执行一个任务。当一个作业被提交到集群时，JobManager 会为该作业分配任务，并将任务分配给空闲的 TaskManager 的任务槽。当一个任务槽被占用时，它将不再可用，直到任务完成并释放该槽。

当 Flink 程序提交后，会创建一个客户端（Client）来进行预处理（上传 JAR 文件和相关配置文件），并根据程序代码生成数据流图（DataFlow Graph），然后对其进行优化，生成作业图（Job Graph），并通过执行系统将作业提交到 JobManager 进行调度。

在实现上，Flink 基于 Actor 模型实现了 JobManager 和 TaskManager，JobManager 与 TaskManager 之间的信息交换，都是通过事件的方式来进行的。Actor 模型属于并发组件模型，封装了其本身的状态和行为，在进行并发编程时，Actor 模型只需要关注消息和其本身，从而避免了并发环境下的锁和内存原子性等问题。

Flink 系统主要包含如下 3 个进程。

（1）JobManager。JobManager 是 Flink 系统的协调者，是管理节点（Master）。它负责接收 Flink 作业，并调度组成作业的多个任务的执行。同时，JobManager 还负责收集作业的状态信息，并管理 Flink 集群中的从节点 TaskManager。

（2）TaskManager。TaskManager 是实际负责执行计算的工作节点（Worker），能够执行 Flink 作业的一组任务。每个 TaskManager 负责管理其所在节点上的资源信息，如内存、磁盘、网络，在启动的时候将资源的状态向 JobManager 汇报。

（3）Client。当用户提交一个 Flink 程序时，首先会创建一个 Client。该 Client 会对用户提交的 Flink 程序进行预处理，生成作业图，并将作业图生成的作业提交到 Flink 集群中进行处理，所以 Client 需要从用户提交的 Flink 程序配置中获取 JobManager 的地址，并建立与 JobManager 的连接，将 Flink 作业提交给 JobManager。

Flink 支持多种部署模式，常用的几种模式如下。

（1）Local。本地单机模式，学习测试时使用。

（2）Standalone。独立集群模式，Flink 自带集群，开发、测试环境中使用。

（3）HA。独立集群高可用模式，Flink 自带集群，开发、测试及生产环境中使用。

（4）YARN。计算资源统一由 Hadoop YARN 管理，生产环境中使用。

在实际开发中，考虑到 YARN 的资源可以按需使用，能够提高集群的资源利用率，且任务根据 YARN 优先级运行作业。此外，基于 YARN 调度系统，Flink 能够自动地处理各个角色的容错（Failover），因此部署 Flink 时更多地会使用 YARN 模式。YARN 模式运行时有如下特点。

① JobManager 进程和 TaskManager 进程都由 Yarn NodeManager（负责计算节点的资源分配和任务管理）监控。

② 如果 JobManager 进程异常退出，则 Yarn ResourceManager（负责整个集群的资源分

配和管理)会将 JobManager 重新调度到其他机器。

③ 如果 TaskManager 进程异常退出,则 JobManager 会收到消息并重新向 Yarn ResourceManager 申请资源,重新启动 TaskManager。

下面将详细介绍 Flink 的安装与配置。

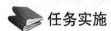**任务实施**

微课 5-5　Flink 的安装与使用

Flink 安装与配置的主要过程如下。

(1)解压 Flink 安装包并重命名文件

操作命令如下。

```
[root@master1 opt]# tar -zxvf flink-1.10.0-bin-scala_2.11.tgz -C /usr/local
[root@master1 local]# mv flink-1.10.0 flink
```

(2)配置环境变量

① 编辑 ~/.bashrc 配置文件,编辑内容如下。

```
export HADOOP_CLASSPATH=`hadoop classpath`
export FLINK_HOME=/usr/local/flink
export HADOOP_CONF_DIR=/usr/local/hadoop/etc/hadoop
PATH=$PATH:$FLINK_HOME/bin
```

② 修改完成后,按"Esc"键返回命令行模式,通过输入":wq!"保存修改并退出,随后执行"source ~/.bashrc"命令,使配置文件立即生效,操作命令如下。

```
[root@master1 local]$ source ~/.bashrc
```

(3)配置 yarn-site.xml 文件

① yarn-site.xml 文件配置内容如下。

```
<configuration>
<!-- Site specific YARN configuration properties -->
<property>
<name>yarn.nodemanager.aux-services</name>
<value>mapreduce_shuffle</value>
</property>
<property>
<name>yarn.nodemanager.pmem-check-enabled</name>
<value>false</value>
</property>
<property>
<name>yarn.nodemanager.vmem-check-enabled</name>
<value>false</value>
</property>
<property>
<name>yarn.resourcemanager.hostname</name>
<value>master1</value>
</property>
```

```
<property>
<name>yarn.resourcemanager.webapp.address</name>
<value>master1:8088</value>
</property>
</configuration>
```

② 启动 Hadoop 集群，操作命令及结果如下。

```
[root@master1 hadoop]# start-all.sh
This script is Deprecated. Instead use start-dfs.sh and start-yarn.sh
Starting namenodes on [master1]
master1: starting namenode, logging to /usr/local/hadoop/logs/hadoop-root-namenode-master1.out
slave1: starting datanode, logging to /usr/local/hadoop/logs/hadoop-root-datanode-slave1.out
slave2: starting datanode, logging to /usr/local/hadoop/logs/hadoop-root-datanode-slave2.out
Starting secondary namenodes [0.0.0.0]
0.0.0.0: starting secondarynamenode, logging to /usr/local/hadoop/logs/hadoop-root-secondarynamenode-master1.out
starting yarn daemons
starting resourcemanager, logging to /usr/local/hadoop/logs/yarn-root-resourcemanager-master1.out
slave2: starting nodemanager, logging to /usr/local/hadoop/logs/yarn-root-nodemanager-slave2.out
slave1: starting nodemanager, logging to /usr/local/hadoop/logs/yarn-root-nodemanager-slave1.out
```

③ 在 master1、slave1、slave2 节点上查看进程。因为是 Flink On Yarn 模式，所以注意查看 ResourceManager 和 NodeManager 进程是否正确启动。

在 master1 节点上查看进程，操作命令及结果如下。

```
[root@master1 hadoop]# jps
1763 NameNode
2152 ResourceManager
2414 Jps
1983 SecondaryNameNode
```

在 slave1 节点上查看进程，操作命令及结果如下。

```
[root@slave1 ~]# jps
832 DataNode
945 NodeManager
1432 Jps
```

在 slave2 节点上查看进程，操作命令及结果如下。

```
[root@slave2 ~]# jps
369 Jps
```

```
121 DataNode
234 NodeManager
```

（4）运行 Flink 自带的 WordCount 应用进行环境测试

程序正确运行即代表 Flink 配置正确，操作命令及结果如下。

```
[root@master1 hadoop]# flink run -m yarn-cluster /usr/local/flink/examples/
batch/WordCount.jar
SLF4J: Class path contains multiple SLF4J bindings.
#省略部分信息
Program execution finished
Job with JobID c9898c7e66dbe11ef6b38f8a2c617703 has finished.
Job Runtime: 8336 ms
Accumulator Results:
- b4c3e039a81dabc272adcd362ab3af18 (java.util.ArrayList) [170 elements]
(a,5)
(action,1)
(after,1)
(against,1)
(all,2)
(and,12)
(arms,1)
(arrows,1)
(awry,1)
(ay,1)
(bare,1)
```

项目小结

本项目介绍了 Sqoop、Pig、Flume、Kafka 和 Flink 的相关知识，完成了 Sqoop、Pig、Flume、Kafka 和 Flink 的安装与配置等操作；重点介绍了使用 Sqoop 完成 MySQL 和 HDFS 之间的数据互导，使用 Pig 进行简单的数据分析，使用 Flume 完成日志数据上传 HDFS 和运行 Flink 自带的 WordCount 应用进行环境测试等基本操作。学习完本项目，读者可对 Hadoop 部分生态组件的相关技术与应用建立清晰的认识，为今后使用 Flume、Kafka 和 Flink 等构建大数据实时分析日志系统奠定良好的基础。

课后练习

1. 如何解决 Sqoop 导入、导出 null 存储一致性问题？
2. 为什么在 Pig 编程时需要 MapReduce？

项目 5　Hadoop 生态组件的安装与使用

3. 请简述 Flume 的组成。
4. 为什么要使用 Kafka？
5. Kafka 中的 ZooKeeper 起到了什么作用？
6. Flink 数据倾斜如何查看？Flink 出现数据倾斜时应如何处理？
7. Flink 提交时并行度如何确定？资源如何配置？

项目 ❻ Hadoop HA 集群搭建

学习目标

【知识目标】
理解传统 Hadoop 集群与 Hadoop HA 集群的区别。
熟悉 Hadoop HA 集群实现的原理。
熟悉 NameNode 和 DataNode 等服务的启动命令。

【素质目标】
具有严谨细致的工作态度和工作作风。
具有良好的团队协作意识和业务沟通能力。
具有良好的表达能力和文档查阅能力。

【技能目标】
学会 Hadoop HA 集群的安装与配置。
学会 Hadoop HA 集群的启动与自动故障转移测试。

项目描述

HA 是 High Availability 的缩写,即高可用。正在工作的机器宕机后,Hadoop HA 集群会自动处理这个异常,并将工作无缝转移到其他备用机器上,以保证服务的高可用。在 Hadoop 1.x 集群中,由于只有一个 NameNode,所以有可能产生单点故障(Single Point of Failure,SPOF)。为了解决 Hadoop 1.x 集群存在的单点故障,从 Hadoop 2.x 集群开始,集群中允许存在 2 个 NameNode,一个处于活跃状态,另一个处于待命状态,处于活跃状态的 NameNode 可对外提供服务。

本项目主要完成 Hadoop HA 集群的搭建,实现 Hadoop 集群的高可用,并完成 Hadoop HA 集群的启动与自动故障转移测试。

任务 6.1　Hadoop HA 集群环境搭建

任务描述

(1)借助学习论坛、网络视频等网络资源和各种图书资源,熟悉 NameNode 在 Hadoop 集群中的作用。

(2)熟悉 Hadoop HA 集群环境的搭建。

任务目标

学会 Hadoop HA 集群的安装与配置。

项目 6 Hadoop HA 集群搭建

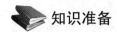 知识准备

在 Hadoop 2.x 之前，NameNode 在 HDFS 集群中是一个非常重要的组件，在分布式集群环境中有可能出现 NameNode 崩溃或其他意外情况。NameNode 影响 HDFS 集群使用的原因主要有以下两种。

（1）发生意外事件。如机器崩溃，导致集群不可使用，直到管理员重启机器才能恢复。

（2）计划内的维护事件。如 NameNode 机器上的软件或硬件升级，导致集群无法使用。

每个集群都至少有一个 NameNode，如果该机器或进程不可用，则将导致整个集群不可用，直到 NameNode 被重新启动或在另一台机器上被恢复。

HDFS 集群的高可用性解决了上述问题，它提供了在同一集群中以主动/被动配置运行两个冗余的 NameNode 的选项，并有一个热备用。这允许集群在一台机器崩溃的情况下快速切换到一个新的 NameNode，或者在计划维护的情况下由管理员发起主动的故障转移，以保证整个集群的正常运行。

为了让待机节点与主动节点保持状态同步，现有的实现方法是使两个节点共享存储设备上的目录。

共享目录的实现有以下两种方式。

（1）使用仲裁日志管理器（Quorum Journal Manager，QJM）。

（2）使用网络文件系统（Network File System，NFS）。

当主动节点执行任何命名空间的修改时，它会将修改记录持久地记录到共享目录的 Edit Logs 中。待机节点会一直监视这个目录的修改情况，当它监视到目录被修改时，就会把这些修改操作同步到自己的命名空间。在需要进行故障转移时，待机节点将确保它在将自己提升到活跃状态之前，已经从共享目录中读取了所有的修改内容，从而确保在进行故障转移时，命名空间的状态是完全同步的。

为了提供快速的故障转移服务，待机节点需要拥有关于集群中数据块位置的最新信息。为了实现这一点，DataNode 中会存储所有 NameNode 的位置信息，并向所有 NameNode 发送数据块位置信息和心跳信号。

确保 Hadoop HA 集群操作的准确性的关键在于保证处于活跃状态的 NameNode 是唯一的，否则，命名空间的状态会因双节点并发操作而迅速分化，即"脑裂（Split Brain）"，从而引发数据丢失或更严重后果。因此，为防止发生"脑裂"，管理员必须为共享目录配置至少一种围栏方法（Fencing Method）。在故障转移过程中，如果不能验证前一个处于活跃状态的节点已经放弃了其活跃状态，围栏方法将负责切断前一个处于活跃状态的节点对共享目录的访问，以防止其对命名空间进行进一步的编辑，从而保证新的处于活跃状态的节点能够安全地进行故障转移。

自动故障转移在 HDFS 部署中增加了两个新的组件：ZooKeeper Quorum 和 ZKFailover Controller（简称 ZKFC）。

ZooKeeper Quorum 在分布式系统中起着关键作用，它能够确保系统的高可用性和一致性。在一个由多个节点组成的 ZooKeeper 集群中，ZooKeeper Quorum 的作用如下。

（1）确保选举的唯一性与可靠性。ZooKeeper Quorum 通过要求多数节点（超过半数）同意来选举主动节点及确认数据更新，有效避免了"脑裂"问题，即使面对网络分割或节点故

障，仍能维持服务连续性和一致性，为分布式系统提供了稳定可靠的基础协调服务。

（2）保证数据的安全性与一致性。ZooKeeper Quorum 确保了数据在被多数节点确认并持久化后，才会通知客户端写操作成功。在此过程中，尽管不是所有节点都立即参与数据的初次复制，但集群通过持续的同步机制，最终会在所有节点上保持数据的一致性。

ZKFC 是一个新的组件，它是 ZooKeeper 的客户端，也负责监视和管理 NameNode 的状态。每台运行 NameNode 的机器都运行一个 ZKFC。ZKFC 的功能如下。

（1）健康监测。ZKFC 定期用健康检查命令"ping"检查其本地 NameNode 的状态。只要 NameNode 及时响应健康状态，ZKFC 就认为该节点是健康的。如果该节点崩溃、冻结或以其他方式进入不健康状态，则 ZKFC 会将它标记为不健康。

（2）ZooKeeper 会话管理。当本地 NameNode 健康时，ZKFC 会在 ZooKeeper 中保持一个开放会话。如果本地 NameNode 处于活跃状态，则它还持有一个特殊的锁节点，即上文提到的独占锁。当本地 NameNode 宕机时，其锁节点将会被删除，处于待命状态的 NameNode 得到锁节点，并升级为主动节点，同时标记状态为活跃。当宕机的 NameNode 重新启动后，它会再次注册 ZooKeeper，由于已有锁节点，因此该 NameNode 自动变为待命状态。

（3）基于 ZooKeeper 的选举。如果本地 NameNode 是健康的，且目前没有其他节点持有锁节点，则它将尝试获得该锁节点。如果其成功了，那么其"赢得了选举"，并负责进行故障转移，使其本地 NameNode 处于活跃状态。故障转移过程的原则如下：首先，对原主动节点采取围栏措施，防止其在非正常状态下操作；其次，将待机节点激活为新的主动节点，以确保服务连续性和命名空间的统一管理。

 任务实施

1. 配置静态 IP 地址

在宿主机中，创建自定义网桥，指定网段、网关，操作命令及结果如下。

```
[root@CentOS ~]# docker network create --subnet=172.32.0.0/24
--gateway=172.32.0.1 HA
f1bad6fa92064749b7e9df80ae3832fd0a33ba644b781c83b2d7da6510ea5723
```

微课 6-1　Hadoop HA 集群环境搭建、启动与自动故障转移测试

2. 为 Docker 容器添加端口映射

创建 3 个 Docker 容器，并添加端口映射，操作命令如下。

```
[root@CentOS ~]# docker run -dit -h master1 --name master1 --network=HA
--ip=172.32.0.20 -p 10022:22 -p 50070:50070 -p 8088:8088 -p 8042:8042 -p 2181:2181
--privileged=true centos:7 init


[root@CentOS ~]# docker run -dit -h slave1 --name slave1 --network=HA
--ip=172.32.0.21 -p 10023:22 -p 50071:50070
-p 8089:8088 -p 2182:2181 -p 8043:8042 --privileged=true centos:7 init
```

```
[root@CentOS ~]# docker run -dit -h slave2 --name slave2 --network=HA
--ip=172.32.0.22 -p 10024:22 -p 8090:8088 -p 2183:2181 -p 8044:8042
--privileged=true centos:7 init
```

3. 将安装包复制到 master1 容器

操作命令及结果如下。

```
[root@CentOS opt]# docker cp /opt/hadoop-2.7.7.tar.gz master1:/opt
Successfully copied 218.7MB to master1:/opt
[root@CentOS opt]# docker cp /opt/zookeeper-3.4.8.tar.gz master1:/opt
Successfully copied 22.26MB to master1:/opt
[root@CentOS opt]# docker cp /opt/jdk-8u341-linux-x64.tar.gz master1:/opt
Successfully copied 148.2MB to master1:/opt
```

4. 安装并启用 SSH 服务

进入 master1 容器，安装并启用 SSH 服务，操作命令如下。

```
[root@CentOS opt]# docker exec -it master1 bash
[root@master1 /]# yum install -y net-tools vim openssh openssh-clients openssh-server which
[root@master1 /]# systemctl start sshd
```

5. 解压并重命名文件

将 JDK、Hadoop、ZooKeeper 文件解压到指定目录中，并重命名文件，操作命令如下。

```
[root@master1 opt]# tar -zxvf jdk-8u341-linux-x64.tar.gz -C /usr/lib/
[root@master1 opt]# tar -zxvf hadoop-2.7.7.tar.gz -C /usr/local/
[root@master1 opt]# tar -zxvf zookeeper-3.4.8.tar.gz -C /usr/local/
[root@master1 opt]# cd /usr/lib
[root@master1 lib]# mv jdk1.8.0_341 jdk1.8.0
[root@master1 lib]# cd /usr/local/
[root@master1 local]# mv hadoop-2.7.7 hadoop
[root@master1 local]# mv zookeeper-3.4.8/ zookeeper
```

6. 添加环境变量

编辑 ~/.bashrc 文件，将相应软件的安装路径添加到环境变量中，编辑内容如下。

```
export JAVA_HOME=/usr/lib/jdk1.8.0
export HADOOP_HOME=/usr/local/hadoop
export ZOOKEEPER_HOME=/usr/local/zookeeper
PATH=$PATH:$JAVA_HOME/bin:$HADOOP_HOME/bin:$HADOOP_HOME/sbin:$ZOOKEEPER_HOME/bin
```

7. 配置免密登录

生成密钥，配置免密登录并进行测试，操作命令及结果如下。

```
[root@master1 ~]# ssh-keygen -t rsa
[root@master1 ~]# ssh-copy-id master1
```

```
[root@master1 ~]# ssh-copy-id slave1
[root@master1 ~]# ssh-copy-id slave2
[root@master1 ~]# ssh slave1
Last login: Mon Mar  6 08:15:56 2023 from 192.168.18.1
```

8. 配置 ZooKeeper

① 进入 zookeeper/conf 目录，通过复制 zoo_sample.cfg 文件创建 zoo.cfg 文件，操作命令如下。

```
[root@master1 conf]# cp zoo_sample.cfg zoo.cfg
```

② 编辑 zoo.cfg 文件，操作命令及修改文件内容如下。

```
[root@master1 conf]# vi zoo.cfg
tickTime=2000
initLimit=10
syncLimit=5
clientPort=2181
dataDir=/usr/local/zookeeper/data
dataLogDir=/usr/local/zookeeper/logs
server.1=master1:2888:3888
server.2=slave1:2889:3889
server.3=slave2:2890:3890
```

其中，"dataDir"和"dataLogDir"配置项中的目录需要手动创建，操作命令如下。

```
[root@master1 zookeeper]# mkdir data
[root@master1 zookeeper]# mkdir logs
```

③ 进入新创建的 data 目录，创建 myid 文件，操作命令如下。

```
[root@master1 zookeeper]# cd data/
[root@master1 data]# touch myid
```

9. 配置 Hadoop

① 配置完 ZooKeeper 后，切换到/usr/local/hadoop/etc/hadoop 目录，查看目录内容，操作命令及结果如下。

```
[root@master1 hadoop]# ls
capacity-scheduler.xml        httpfs-env.sh                 mapred-env.sh
configuration.xsl             httpfs-log4j.properties       mapred-queues.xml.template
container-executor.cfg        httpfs-signature.secret       mapred-site.xml.template
core-site.xml                 httpfs-site.xml               slaves
hadoop-env.cmd                kms-acls.xml                  ssl-client.xml.example
hadoop-env.sh                 kms-env.sh                    ssl-server.xml.example
hadoop-metrics.properties     kms-log4j.properties          yarn-env.cmd
hadoop-metrics2.properties    kms-site.xml                  yarn-env.sh
hadoop-policy.xml             log4j.properties              yarn-site.xml
hdfs-site.xml                 mapred-env.cmd
```

项目6 Hadoop HA 集群搭建

② 编辑 hadoop-env.sh 文件，修改"JAVA_HOME"配置项，修改内容如下。

```
# The java implementation to use.
export JAVA_HOME=/usr/lib/jdk1.8.0
```

③ 在/usr/local/hadoop/etc/hadoop 目录中，通过复制 mapred-site.xml.template 文件生成 mapred-site.xml 文件，操作命令如下。

```
[root@master1 hadoop]# cp mapred-site.xml.template mapred-site.xml
```

④ 分别编辑 core-site.xml、hdfs-site.xml、mapred-site.xml、yarn-site.xml 这 4 个文件。

编辑 core-site.xml 文件，文件内容如下。

```
<configuration>

<!-- 决定是使用 fs.default.name 还是使用 fs.defaultFS，首先判断是否开启了 NameNode 的高可
用功能，如果开启了则使用 fs.defaultFS，如果未开启则使用 fs.default.name -->
<!--指定默认的 HDFS 路径 -->
    <property>
        <name>fs.defaultFS</name>
        <value>hdfs://mycluster</value>
    </property>
<!--指定 NameNode、DataNode、SecondaryNameNode 等存储数据的公共目录-->
    <property>
        <name>hadoop.tmp.dir</name>
        <value>/usr/local/hadoop/tmp</value>
    </property>
<!-- 指定 ZooKeeper 地址-->
    <property>
        <name>ha.zookeeper.quorum</name>
        <value>master1:2181,slave1:2181,slave2:2181</value>
    </property>
<!-- 设置 Hadoop 连接 ZooKeeper 的超时时间-->
    <property>
        <name>ha.zookeeper.session-timeout.ms</name>
        <value>30000</value>
    </property>
<!-- 设置垃圾回收时间，以分钟为单位，如设置为 1440min，刚好是 1 天-->
    <property>
        <name>fs.trash.interval</name>
        <value>1440</value>
    </property>
</configuration>
```

编辑 hdfs-site.xml 文件，文件内容如下。

```
<configuration>
<!-- 指定 JournalNode 集群之间通信的超时时间 -->
    <property>
```

```xml
            <name>dfs.qjournal.start-segment.timeout.ms</name>
            <value>60000</value>
    </property>
<!--指定 NameNode 的组名称，可以自定义 -->
    <property>
            <name>dfs.nameservices</name>
            <value>mycluster</value>
    </property>
<!--指定 NameNode 组 mycluster 的成员-->
    <property>
            <name>dfs.ha.namenodes.mycluster</name>
            <value>master1,slave1</value>
    </property>
<!--配置 master1 的 RPC 地址及端口号-->
    <property>
            <name>dfs.namenode.rpc-address.mycluster.master1</name>
            <value>master1:9000</value>
    </property>
<!--配置 slave1 的 RPC 地址及端口号-->
    <property>
            <name>dfs.namenode.rpc-address.mycluster.slave1</name>
            <value>slave1:9000</value>
    </property>
<!--配置 master1 的 HTTP 地址及端口号-->
    <property>
            <name>dfs.namenode.http-address.mycluster.master1</name>
            <value>master1:50070</value>
    </property>
<!--配置 slave1 的 HTTP 地址及端口号-->
    <property>
            <name>dfs.namenode.http-address.mycluster.slave1</name>
            <value>slave1:50070</value>
    </property>
<!-- 指定 NameNode 的 edits 元数据在 JournalNode 上的存储位置 -->
    <property>
            <name>dfs.namenode.shared.edits.dir</name>
            <value>qjournal://master1:8485;slave1:8485;slave2:8485/mycluster
            </value>
    </property>
<!-- 当集群出现故障时，指定负责执行故障转移的对象 -->
    <property>
            <name>dfs.client.failover.proxy.provider.mycluster</name>
```

```xml
        <value>
    org.apache.hadoop.hdfs.server.namenode.ha.ConfiguredFailoverProxyProvider
        </value>
    </property>
<!-- 配置隔离机制,多个机制用换行分隔,即每个机制占用一行-->
    <property>
        <name>dfs.ha.fencing.methods</name>
        <value>
        sshfence
        shell(/bin/true)
        </value>
    </property>
<!-- 如果为"true",则在HDFS中启用权限检查;如果为"false",则关闭权限检查;默认值为
"true" -->
    <property>
        <name>dfs.permissions.enabled</name>
        <value>false</value>
    </property>

<!-- 指定是否支持文件追加,默认为"false",表示不支持-->
    <property>
        <name>dfs.support.append</name>
        <value>true</value>
    </property>
<!-- 使用sshfence隔离机制时需要配置SSH免登录 -->
    <property>
        <name>dfs.ha.fencing.ssh.private-key-files</name>
        <value>/root/.ssh/id_rsa</value>
    </property>
<!--指定数据备份的个数 -->
    <property>
        <name>dfs.replication</name>
        <value>2</value>
    </property>
<!-- 指定HDFS元数据存储的路径 -->
    <property>
        <name>dfs.namenode.name.dir</name>
        <value>/usr/local/hadoop/dfs/name</value>
    </property>
<!-- 指定HDFS存储数据的路径 -->
    <property>
        <name>dfs.datanode.data.dir</name>
```

```xml
        <value>/usr/local/hadoop/dfs/data</value>
    </property>
<!-- 指定JournalNode存储日志文件的路径 -->
    <property>
        <name>dfs.journalnode.edits.dir</name>
        <value>/usr/local/hadoop/tmp/journal</value>
    </property>
<!-- 开启自动故障转移 -->
    <property>
        <name>dfs.ha.automatic-failover.enabled</name>
        <value>true</value>
    </property>
<!-- 开启WebHDFS功能（基于REST的接口服务） -->
    <property>
        <name>dfs.webhdfs.enabled</name>
        <value>true</value>
    </property>
<!-- 配置sshfence隔离机制超时时间 -->
    <property>
        <name>dfs.ha.fencing.ssh.connect-timeout</name>
        <value>30000</value>
    </property>
<!-- 设定从CLI手动运行的故障转移功能等待健康检查、服务状态的超时时间 -->
    <property>
        <name>ha.failover-controller.cli-check.rpc-timeout.ms</name>
        <value>60000</value>
    </property>
</configuration>
```

编辑mapred-site.xml文件，文件内容如下。

```xml
<configuration>
<!-- 设定MapReduce执行框架为YARN-->
    <property>
        <name>mapreduce.framework.name</name>
        <value>yarn</value>
    </property>
<!-- 设定JobHistory的IPC地址-->
    <property>
        <name>mapreduce.jobhistory.address</name>
        <value>master1:10020</value>
    </property>
<!-- 设定JobHistory的Web地址-->
    <property>
```

```xml
        <name>mapreduce.jobhistory.webapp.address</name>
        <value>master1:19888</value>
    </property>
</configuration>
```

编辑 yarn-site.xml 文件，文件内容如下。

```xml
<configuration>
<!-- 开启 ResourceManager 高可用功能 -->
    <property>
        <name>yarn.resourcemanager.ha.enabled</name>
        <value>true</value>
    </property>
    <!-- 指定 ResourceManager 的集群 ID-->
    <property>
         <name>yarn.resourcemanager.cluster-id</name>
         <value>yrc</value>
    </property>
<!-- 指定 ResourceManager 的名称-->
    <property>
        <name>yarn.resourcemanager.ha.rm-ids</name>
        <value>rm1,rm2</value>
    </property>
<!-- 指定 master1 节点的 ResourceManager 的主机名-->
    <property>
        <name>yarn.resourcemanager.hostname.rm1</name>
        <value>master1</value>
    </property>
<!-- 指定 slave1 节点的 ResourceManager 的主机名-->
    <property>
        <name>yarn.resourcemanager.hostname.rm2</name>
        <value>slave1</value>
    </property>
<!-- 指定 ResourceManager 使用 ZK 集群地址-->
    <property>
        <name>yarn.resourcemanager.zk-address</name>
        <value>master1:2181,slave1:2181,slave2:2181</value>
    </property>
<!-- 配置 YARN 运行 shuffle 的方式，选择 MapReduce 的默认 shuffle 算法-->
    <property>
        <name>yarn.nodemanager.aux-services</name>
        <value>mapreduce_shuffle</value>
    </property>
<!--是否开启日志聚集功能-->
```

```xml
    <property>
            <name>yarn.log-aggregation-enable</name>
            <value>true</value>
    </property>
<!--配置聚集的日志在HDFS上的最长保存时间-->
    <property>
            <name>yarn.log-aggregation.retain-seconds</name>
            <value>86400</value>
    </property>
<!-- 开启Recovery（重启作业保留机制）。开启后，ResourceManager会将应用的状态等信息保存到
yarn.resourcemanager.store.class配置的存储介质中，重启后会加载这些信息，并且NodeManager
会将还在运行的Container（YARN中资源的抽象，封装某个节点上一定量的资源，如CPU和内存资源）
信息同步到ResourceManager，整个过程不影响作业的正常运行-->
    <property>
            <name>yarn.resourcemanager.recovery.enabled</name>
            <value>true</value>
    </property>
<!-- 指定yarn.resourcemanager.store.class的存储介质（HA集群只支持ZKRMStateStore）
-->
    <property>
            <name>yarn.resourcemanager.store.class</name>
            <value>org.apache.hadoop.yarn.server.resourcemanager.recovery.ZKRMStateStore</value>
    </property>
</configuration>
```

⑤ 修改slaves文件（3.x版本为workers），编辑内容如下。

```
master1
slave1
slave2
```

⑥ 在core-site.xml、hdfs-site.xml、mapred-site.xml、yarn-site.xml这4个配置文件中，所涉及的配置项中的目录需要手动创建，操作命令如下。

```
[root@master1 hadoop]# mkdir logs
[root@master1 hadoop]# mkdir dfs
[root@master1 hadoop]# mkdir tmp
[root@master1 hadoop]# cd dfs
[root@master1 dfs]# mkdir data
[root@master1 dfs]# mkdir name
[root@master1 dfs]# cd ..
[root@master1 hadoop]# cd tmp/
[root@master1 tmp]# mkdir journal
```

⑦ 在master1节点上将.bashrc、jdk1.8.0、hadoop和zookeeper文件分别复制到slave1及

slave2 节点上,操作命令如下。

```
[root@master1 ~]# scp -r ~/.bashrc slave1:~/.bashrc
[root@master1 ~]# scp -r ~/.bashrc slave2:~/.bashrc
[root@master1 ~]# scp -r /usr/lib/jdk1.8.0 slave1:/usr/lib/
[root@master1 ~]# scp -r /usr/lib/jdk1.8.0 slave2:/usr/lib/
[root@master1 ~]# scp -r /usr/local/zookeeper slave1:/usr/local/
[root@master1 ~]# scp -r /usr/local/zookeeper slave2:/usr/local/
[root@master1 ~]# scp -r /usr/local/hadoop slave1:/usr/local/
[root@master1 ~]# scp -r /usr/local/hadoop slave2:/usr/local/
```

⑧ 分别在 slave1 和 slave2 节点上执行"source ~/.bashrc"命令,使环境变量立即生效。

10. 启动 Hadoop 集群

(1)启动 ZooKeeper 集群并查看其状态

在 master1 节点上启动 ZooKeeper 集群并查看其状态,操作命令及结果如下。

```
[root@master1 bin]# ./zkServer.sh start
ZooKeeper JMX enabled by default
Using config: /usr/local/zookeeper/bin/../conf/zoo.cfg
Starting zookeeper ... STARTED
[root@master1 bin]# ./zkServer.sh status
ZooKeeper JMX enabled by default
Using config: /usr/local/zookeeper/bin/../conf/zoo.cfg
Mode: follower
```

在 slave1 节点上启动 ZooKeeper 集群并查看其状态,操作命令及结果如下。

```
[root@slave1 zookeeper]# cd bin
[root@slave1 bin]# ./zkServer.sh start
ZooKeeper JMX enabled by default
Using config: /usr/local/zookeeper/bin/../conf/zoo.cfg
Starting zookeeper ... STARTED
[root@slave1 bin]# ./zkServer.sh status
ZooKeeper JMX enabled by default
Using config: /usr/local/zookeeper/bin/../conf/zoo.cfg
Mode: follower
```

在 slave2 节点上启动 ZooKeeper 集群并查看其状态,操作命令及结果如下。

```
[root@slave2 zookeeper]# cd bin
[root@slave2 bin]# ./zkServer.sh start
ZooKeeper JMX enabled by default
Using config: /usr/local/zookeeper/bin/../conf/zoo.cfg
Starting zookeeper ... STARTED
[root@slave2 bin]# ./zkServer.sh status
```

```
ZooKeeper JMX enabled by default
Using config: /usr/local/zookeeper/bin/../conf/zoo.cfg
Mode: leader
```
（2）初始化 HA 在 ZooKeeper 中的状态

操作命令及部分结果如下。

```
[root@master1 bin]# ./hdfs zkfc -formatZK
23/03/06 14:42:42 INFO ha.ActiveStandbyElector: Successfully created
/hadoop-ha/mycluster in ZK.
23/03/06 14:42:42 INFO ha.ActiveStandbyElector: Session connected.
23/03/06 14:42:42 INFO zookeeper.ZooKeeper: Session: 0x386b7555fb80000 closed
23/03/06 14:42:42 INFO zookeeper.ClientCnxn: EventThread shut down
```
（3）启动 JournalNode

① 分别在 master1、slave1 和 slave2 节点上启动 JournalNode，操作命令如下。

```
[root@master1 sbin]# ./hadoop-daemon.sh start journalnode
[root@slave1 sbin]# ./hadoop-daemon.sh start journalnode
[root@slave2 sbin]# ./hadoop-daemon.sh start journalnode
```

② 在 master1 节点上初始化 NameNode，如果看到返回信息中包含 "status 0"，则代表初始化成功，操作命令及部分结果如下。

```
[root@master1 bin]# hdfs namenode -format
23/03/06 14:47:20 INFO util.ExitUtil: Exiting with status 0
23/03/06 14:47:20 INFO namenode.NameNode: SHUTDOWN_MSG:
/************************************************************
SHUTDOWN_MSG: Shutting down NameNode at master1/192.168.18.110
************************************************************/
```

（4）启动 Hadoop 并查看节点的进程信息

在 master1 节点上执行 "start-all.sh" 命令，并分别查看 master1、slave1 和 slave2 节点的进程信息，操作命令及结果如下。

```
[root@master1 hadoop]# start-all.sh
[root@master1 hadoop]# jps
4896 ResourceManager
225 QuorumPeerMain
5330 Jps
3879 JournalNode
4072 DFSZKFailoverController
5016 NodeManager
3673 DataNode
3565 NameNode

[root@slave1 hadoop]# jps
2322 Jps
1783 DFSZKFailoverController
```

```
218 QuorumPeerMain
1596 DataNode
1692 JournalNode
2190 NodeManager

[root@slave2 hadoop]# jps
228 QuorumPeerMain
1173 DataNode
1269 JournalNode
1592 NodeManager
1724 Jps
```

（5）格式化主、从节点

将 master1 节点元数据复制到 slave1 和 slave2 节点上，因为之前 NameNode、DataNode 的数据存储在 hadoop/dfs 目录下，JournalNode 的数据存储在 hadoop/tmp 目录下，所以直接将 dfs 和 tmp 目录复制到 slave1 和 slave2 节点上，操作命令如下。

```
[root@master1 hadoop]# scp -r /usr/local/hadoop/tmp/* slave1:/usr/local/hadoop/tmp/
[root@master1 hadoop]# scp -r /usr/local/hadoop/dfs/* slave1:/usr/local/hadoop/dfs/
[root@master1 hadoop]# scp -r /usr/local/hadoop/tmp/* slave2:/usr/local/hadoop/tmp/
[root@master1 hadoop]# scp -r /usr/local/hadoop/dfs/* slave2:/usr/local/hadoop/dfs/
```

（6）在 slave1 节点上启动 ResourceManager 和 NameNode

操作命令及结果如下。

```
[root@slave1 sbin]# ./yarn-daemon.sh start resourcemanager
[root@slave1 sbin]# ./hadoop-daemon.sh start namenode
[root@slave1 sbin]# jps
2646 Jps
1783 DFSZKFailoverController
2439 NameNode
218 QuorumPeerMain
1596 DataNode
1692 JournalNode
2190 NodeManager
2590 ResourceManage
```

（7）访问 Hadoop 的 Web UI 界面

① 在 Windows 浏览器地址栏中输入 http://宿主机 IP 地址:50070，可以看到 master1 节点的 NameNode 为活跃状态（active），如图 6-1 所示。

② 在 Windows 浏览器地址栏中输入 http://宿主机 IP 地址:50071，可以看到 slave1 节点的 NameNode 为待命状态（standby），如图 6-2 所示。

图 6-1　master1 节点（NameNode 为活跃状态）

图 6-2　slave1 节点（NameNode 为待命状态）

③ 在 Windows 浏览器地址栏中输入 http://宿主机 IP 地址:8088，可以进入 master1 节点的 YARN Web UI 界面，如图 6-3 所示。

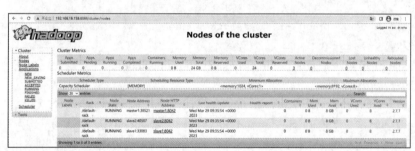

图 6-3　master1 节点的 YARN Web UI 界面

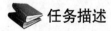　Hadoop HA 集群的启动与自动故障转移测试

📖 任务描述

（1）借助学习论坛、网络视频等网络资源和各种图书资源了解 ZooKeeper 的作用。
（2）熟悉 Hadoop HA 集群的启动与自动故障转移测试操作。

📖 任务目标

（1）学会 NameNode 的格式化操作。
（2）学会 Hadoop HA 集群的启动与自动故障转移测试操作。

项目 6　Hadoop HA 集群搭建

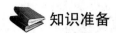 知识准备

在 Hadoop HA 集群中，首先需要对处于活跃状态的 NameNode 进行格式化，Hadoop HA 集群配置完成后，首次启动前需要格式化 HDFS，如果处于活跃状态的 NameNode 格式化失败或者进行过多次格式化，则需要把 NameNode、DataNode、JournalNode（负责维护和管理 NameNode 的编辑日志）的存储目录的内容清空后再进行格式化操作。其次需要对处于待命状态的 NameNode 进行格式化操作。

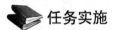 任务实施

1. 查看集群 NameNode 状态

查看集群 NameNode 状态的方式有以下两种。

（1）通过命令查看集群 NameNode 状态

```
hdfs haadmin -getServiceState nn
```

其中，"nn" 是 NameNode 的名称，也就是 NameNode 运行状态为活跃状态和待命状态的节点名称。

查看具体集群的 NameNode 状态，操作命令及结果如下。

```
[root@master1 ~]# hdfs haadmin -getServiceState master1
active
[root@master1 ~]# hdfs haadmin -getServiceState slave1
standby
```

（2）通过 Web 界面查看集群 NameNode 状态

① 在 Windows 浏览器地址栏中输入 "http://宿主机 IP 地址:50070"，查看 master1 节点的 NameNode 状态，如图 6-4 所示，可以看到 master1 节点的 NameNode 为活跃状态。

图 6-4　查看 master1 节点的 NameNode 状态

② 在 Windows 浏览器地址栏中输入 "http://宿主机 IP 地址:50071"，查看 slave1 节点的 NameNode 状态，如图 6-5 所示，可以看到 slave1 节点的 NameNode 为待命状态。

图 6-5 查看 slave1 节点的 NameNode 状态

2. 模拟 NameNode 出现故障并查看 NameNode 的状态

① 由图 6-4 和图 6-5 可知，master1 节点的 NameNode 为活跃状态，slave1 节点的 NameNode 为待命状态，在 master1 节点上执行"hadoop-daemon.sh stop namenode"命令，直接关闭 NameNode，操作命令及结果如下。

```
[root@master1 ~]# hadoop-daemon.sh stop namenode
stopping namenode
```

也可以执行"jps"命令查看 NameNode 进程号，并执行"kill -9 进程号"命令结束 NameNode 进程。

② 刷新网页后，在 Windows 浏览器地址栏中输入"http://宿主机 IP 地址:50070"，发现界面无法访问，如图 6-6 所示。

图 6-6 界面无法访问

此时，在 Windows 浏览器地址栏中输入"http://宿主机 IP 地址:50071"，可以看到 slave1 节点的 NameNode 为活跃状态，如图 6-7 所示。

项目6　Hadoop HA 集群搭建

图 6-7　查看 slave1 节点的 NameNode 状态（master1 节点的 NameNode 出现故障后）

3. 在 master1 节点上启动 NameNode 后再次查看 NameNode 状态

在 master1 节点上执行"hadoop-daemon.sh start namenode"命令，启动 NameNode，依次查看 master1 节点和 slave1 节点上 NameNode 的运行状态，显示结果分别如图 6-8 和图 6-9 所示。

图 6-8　查看 master1 节点的 NameNode 状态（重启 master1 节点的 NameNode 后）

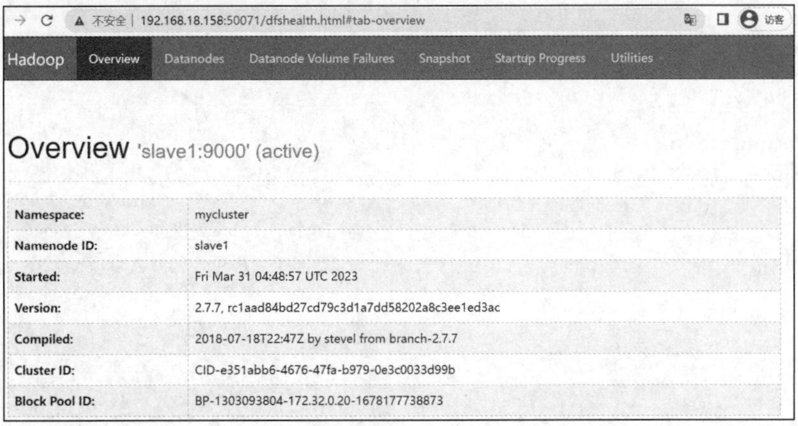

图 6-9　查看 slave1 节点的 NameNode 状态（重启 master1 节点的 NameNode 后）

181

可以发现，当处于活跃状态的 NameNode 因为意外情况停止服务后，原来为待命状态的 NameNode 会自动接替其工作继续提供服务，从而避免 NameNode 的单点故障。

项目小结

本项目介绍了传统 Hadoop 集群与 Hadoop HA 集群的区别，以及 Hadoop HA 实现的原理，并重点介绍了 NameNode 和 DataNode 等服务的启动命令，完成了 Hadoop HA 的安装与配置等操作，以及 Hadoop HA 集群的启动与自动故障转移测试。学习完本项目，读者可对 Hadoop HA 集群自动故障转移机制建立清晰的认知。

课后练习

1. Hadoop HA 如何保证 NameNode 数据的存储安全？
2. Hadoop HA 集群要启动哪些进程，请分别简述它们的作用。
3. 请简述 Hadoop HA 模式下集群启动的步骤。

项目 7 Ambari 搭建与集群管理

学习目标

【知识目标】
了解 Ambari 的功能。
识记 Ambari 的启动命令。

【素质目标】
具有严谨细致的工作态度和工作作风。
具有良好的团队协作意识和业务沟通能力。
具有良好的表达能力和文档查阅能力。

【技能目标】
学会 NTP、本地 yum 源的安装与配置。
学会 Ambari 的安装与配置。
学会通过 Ambari 创建、管理和监控 Hadoop 集群。

项目描述

Apache Ambari 可以通过配置、管理和监控 Hadoop 集群来简化 Hadoop 管理工作。Ambari 提供了一个由 RESTful API 支持的、直观且易用的 Hadoop 管理 Web UI，目前可支持大多数 Hadoop 组件，包括 HDFS、MapReduce、Hive、Pig、HBase、ZooKeeper 和 Sqoop 等。Ambari 使用 Ganglia 收集度量指标，使用 Nagios 支持系统报警。此外，Ambari 支持安装安全的、基于 Kerberos 的 Hadoop 集群，以此实现对 Hadoop 安全的支持，它还提供了基于角色的用户认证、授权和审计功能，并为用户管理集成了轻量目录访问协议（Lightweight Directory Access Protocol，LDAP）和活动目录（Active Directory，AD）。

本项目主要完成 Ambari 平台的搭建，实现使用 Ambari 管理 Hadoop 集群。

任务 7.1 搭建 Ambari 平台

任务描述

（1）学习 Ambari 的相关技术知识，了解其功能。
（2）完成 Ambari 的安装与配置。
（3）利用 Ambari 扩展集群。

任务目标

（1）了解 Hadoop 背景知识，熟悉 Hadoop 的生态系统。
（2）学会 Ambari Server 和 Ambari Agent 的配置方法。

(3)学会利用 Ambari 扩展集群。

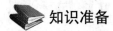 **知识准备**

尽管大数据技术日益普及,但对于很多分布式应用初学者和大数据业务应用开发者而言,学习该技术的门槛始终很高。与大数据有关的多个产品之间的不兼容问题使得产品的快速集成和维护比较困难。不管是 Hadoop 1.x 还是 Hadoop 2.x 的安装,还是 Spark、YARN 等的集成,都不是几行简单的命令就可以完成的,而是需要手工修改很多集群配置,这进一步增加了应用初学者和业务应用开发者的学习及使用难度。然而,有了 Ambari,这些问题即可迎刃而解。

Ambari 管理平台通过安装向导来进行集群的搭建,简化了集群搭建工作。同时,它有一个监控组件——Ambari Metrics,该组件可以提前配置好关键的运维指标,并收集集群中的服务、主机等的运行状态信息,再将其通过 Web 的方式显示出来。用户可以直接查看 Hadoop Core(HDFS 和 MapReduce)及相关项目(如 HBase 和 Hive)是否健康。Ambari 平台的用户界面非常直观,用户可以轻松、有效地查看信息并控制集群。Ambari 支持作业与任务执行的可视化与分析,能够更好地查看依赖和性能。

Ambari 由 Ambari Server 和 Ambari Agent 两部分组成,它们通过心跳机制确认存活状态,其架构如图 7-1 所示。Ambari Server 需要一个数据库来存储元数据,默认使用 PostgreSQL 数据库,也可以使用 MySQL 数据库。Ambari Server 会读取集群中相应服务的配置文件。当用户使用 Ambari 创建集群时,Ambari Server 会将相应的配置文件以及服务生命周期的控制脚本传送到 Ambari Agent 中。Ambari Agent 得到配置文件后,会下载并安装相应的服务,Ambari Server 会通知 Ambari Agent 启动和管理服务。此后,Ambari Server 会定期发送命令到 Ambari Agent 中,以检查服务的状态;Ambari Agent 会将状态信息上报给 Ambari Server,并呈现在 Ambari 的图形用户界面(Graphical User Interface,GUI)上,以方便用户了解集群的各种状态,并进行相应的维护。

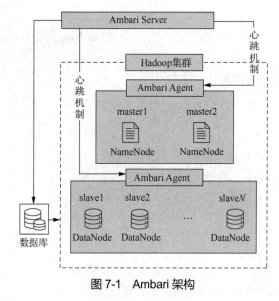

图 7-1　Ambari 架构

项目 7　Ambari 搭建与集群管理

1．Ambari 平台的作用

Ambari 使得系统管理员能够进行以下操作。

（1）搭建 Hadoop 集群

① Ambari 提供了跨任意数量的主机安装 Hadoop 服务的分步向导。

② Ambari 可处理集群的 Hadoop 服务配置。

（2）管理 Hadoop 集群

Ambari 提供对集群的集中管理功能，用于在整个集群中启动、停止和重新配置 Hadoop 服务。

（3）监控 Hadoop 集群

① Ambari 提供了一个仪表板，用于监控 Hadoop 集群的运行状况和状态。

② Ambari 可利用 Ambari 指标系统进行指标收集。

③ Ambari 可利用 Ambari Alert Framework 进行系统警报，并在需要系统管理员注意时（如节点出现故障、剩余磁盘空间不足等）通知系统管理员。

Ambari 使得应用程序开发人员和系统集成商能够使用 Ambari RESTful API 轻松将 Hadoop 的配置、管理和监控功能集成到自己的应用程序中。

2．功能列表

Ambari 提供的主要功能如下。

（1）支持不同级别的操作

① 机器级别的操作（Host Level Action）。

② 组件级别的操作（Component Level Action）。

（2）支持多种用户管理角色

① 集群用户（Cluster User）。能够查看集群和服务的信息，如配置信息、服务状态信息、健康状态信息等。

② 服务操作员（Service Operator）。能够操作服务的生命周期，如启动、停止等，也可以进行 DataNode 数据均衡和 YARN 恢复操作。

③ 服务管理员（Service Administrator）。除了具有服务操作员的功能，还能够进行配置服务、移动 NameNode、启用 HA 等操作。

④ 集群操作员（Cluster Operator）。除了具有服务管理员的功能，还能够对主机（Host）和组件（Component）进行增加、删除等操作。

⑤ 集群管理员（Cluster Administrator）。集群的超级管理员，拥有全部权限，可以操作任意组件。

（3）支持仪表板监控

① 滚动启动（Roll Start）功能。根据服务的依赖关系，按照一定的顺序启动多个服务。

② 关键运维指标监控。能够监控当前组件及整个集群的统计信息。

③ 信息显示。能够显示该服务的模块及其数目信息。界面右上角有关于服务动作（Service Action）的按钮，可进行服务的启动、停止、删除等操作。

④ 提供快速访问链接（Quick Links）。能够链接到组件原生管理界面。

（4）支持告警

① 支持多种告警级别。包括 OK、Warning、Critical、Unknown 和 None。

② 支持多种告警类型。包括 Web、Port、Metric、Aggregate 和 Script。

Ambari 中的告警类型对比如表 7-1 所示。

表 7-1 Ambari 中的告警类型对比

类型	用途	告警级别	阈值是否可配置
Web	用于监测某个 Web UI 地址是否可用	OK、Warning、Critical	否
Port	用来监测机器上的某个端口是否可用	OK、Warning、Critical	是
Metric	用来监测 Metric 相关的配置属性	OK、Warning、Critical	是
Aggregate	用于收集某些告警的状态	OK、Warning、Critical	是
Script	用于监测特定系统或应用的状态	OK、Critical	否

 任务实施

1. Ambari 的安装与配置

（1）基本环境配置

因为 Docker 容器都在同一个网段内直接互通，且配置 hosts 文件后每次重启都会重置，所以不需要对主机名进行额外设置。

① 配置 NTP。在 master 节点上安装网络时间协议（Network Time Protocol，NTP），操作命令如下。

微课 7-1 搭建 Ambari 平台

```
[root@master opt]# yum install -y ntp
```

配置 ntp.conf 文件，操作命令及文件编辑内容如下。

```
[root@master opt]# vi /etc/ntp.conf
//注释或者删除以下 4 行代码
#pool 0.ubuntu.pool.ntp.org iburst
#pool 1.ubuntu.pool.ntp.org iburst
#pool 2.ubuntu.pool.ntp.org iburst
#pool 3.ubuntu.pool.ntp.org iburst
//添加以下两行代码
server 127.127.1.0
fudge 127.127.1.0 stratum 10
[root@master opt]# systemctl enable ntpd
[root@master opt]# systemctl start ntpd
```

在 slave1 节点上安装和配置 NTP，操作命令如下。

项目7　Ambari 搭建与集群管理

```
[root@slave1 opt]# yum install -y ntpdate
[root@slave1 opt]# ntpdate master
[root@slave1 opt]# systemctl enable ntpd
```

② 安装与配置 SSH。安装 SSH 并分别在 master 节点和 slave1 节点上执行 "ssh-keygen -t rsa" 命令生成密钥对，master1 节点的操作命令及部分结果如下。

```
[root@master ~]# yum install -y openssh openssh-clients openssh-server which net-tools
Loaded plugins: fastestmirror, ovl
Determining fastest mirrors
[root@master ~]# ssh-keygen -t rsa
Generating public/private rsa key pair.
Enter file in which to save the key (/root/.ssh/id_rsa):
Enter passphrase (empty for no passphrase):
Enter same passphrase again:
Your identification has been saved in /root/.ssh/id_rsa.
Your public key has been saved in /root/.ssh/id_rsa.pub.
#省略部分信息
```

在 master 节点上分别执行 "ssh-copy-id master" 和 "ssh-copy-id slave1" 命令，分发公钥，完成认证配置操作，操作命令及部分结果如下。

```
[root@master ~]# ssh-copy-id master
#省略部分信息
Are you sure you want to continue connecting (yes/no)? yes
root@master's password:
and check to make sure that only the key(s) you wanted were added.

[root@master ~]# ssh-copy-id slave1
Are you sure you want to continue connecting (yes/no)? yes
root@slave1's password:
```

③ 通过 SSH 登录远程主机，查看能否免密登录 master 节点和 slave1 节点，操作命令及结果如下。

```
[root@master ~]# ssh master
Last login: Wed Mar  8 15:09:42 2023 from 192.168.18.1
[root@master ~]# exit
logout
Connection to master closed.
[root@master ~]# ssh slave1
Last login: Wed Mar  8 15:55:02 2023 from master.ambari
[root@slave1 ~]# exit
logout
Connection to slave1 closed.
[root@master ~]#
```

④ 禁用 SELinux。编辑/etc/selinux/config 配置文件，将配置项修改如下。
`SELINUX=disabled`

⑤ 禁用透明大页（Transparent Huge Page，THP）特性。CentOS 6 引入了 THP，从 CentOS 7 开始，默认启用该特性（操作系统后台会生成一个名为 khugepaged 的进程）。THP 会一直扫描所有进程占用的内存，把 4KB 大小的内存页交换为大内存页。在这个过程中，对操作的内存所进行的各种分配活动都需要各种内存锁，直接影响了程序内存的访问性能，且此过程对于应用是透明的，在应用层面不可控，对于专门优化 4KB 内存页的程序来说，可能会导致其随机性能下降，故需禁用该特性。禁用 THP 的操作命令及结果如下。

```
[root@CentOS ~]# cat /sys/kernel/mm/transparent_hugepage/enabled
[always] madvise never
[root@CentOS ~]# echo never >/sys/kernel/mm/transparent_hugepage/enabled
[root@CentOS ~]# echo never >/sys/kernel/mm/transparent_hugepage/defrag
[root@CentOS ~]# cat /sys/kernel/mm/transparent_hugepage/enabled
always madvise [never]
```

重启计算机后代码会失效，需要再次执行上述命令才能重新禁用 THP。

⑥ 安装与配置 JDK。

在 master 节点上安装与配置 JDK，操作命令及结果如下。

```
[root@master opt]# tar -zxvf jdk-8u181-linux-x64.tar.gz -C /usr/lib
[root@master lib]# cd /usr/lib
[root@master lib]# mv jdk1.8.0_181 jdk1.8.0

[root@master lib]# vim ~/.bashrc
export JAVA_HOME=/usr/lib/jdk1.8.0
PATH=$PATH:$JAVA_HOME/bin

[root@master lib]# source ~/.bashrc

[root@master lib]# java -version
java version "1.8.0_181"
Java(TM) SE Runtime Environment (build 1.8.0_181-b13)
Java HotSpot(TM) 64-Bit Server VM (build 25.181-b13, mixed mode)
```

在 slave1 节点上安装与配置 JDK，操作命令及结果如下。

```
[root@master lib]# scp -r ~/.bashrc slave1: ~/.bashrc
[root@master lib]# scp -r /usr/lib/jdk1.8.0 slave1:/usr/lib
[root@slave1 opt]# source ~/.bashrc
[root@slave1 opt]# java -version
java version "1.8.0_181"
Java(TM) SE Runtime Environment (build 1.8.0_181-b13)
Java HotSpot(TM) 64-Bit Server VM (build 25.181-b13, mixed mode)
```

（2）设置本地仓库

① 安装制作本地仓库的工具。使用"curl"命令来验证网页是否可以访问，操作命令及

项目 7　Ambari 搭建与集群管理

结果如下。

```
[root@master ~]# yum -y install yum-utils createrepo curl
[root@master ~]# yum -y install httpd
[root@master ~]# /sbin/chkconfig httpd on
Note: Forwarding request to 'systemctl enable httpd.service'.
Created symlink from /etc/systemd/system/multi-user.target.wants/httpd.service
to /usr/lib/systemd/system/httpd.service.
[root@master ~]# /sbin/service httpd start
Redirecting to /bin/systemctl start httpd.service
[root@master ~]# curl http://172.22.0.20
```

如果无法访问网页，则通常会报错，如显示 302 或 404 等错误代码。

② 创建 Web 服务目录，操作命令如下。

```
[root@master ~]# mkdir -p /var/www/html/
```

③ 将 ambari-2.6.2.2、HDP-2.6.5.0 和 HDP-GPL-2.6.5.0 等版本的压缩包解压到 /var/www/html 目录下，将 HDP-UTILS 目录通过 "mv" 命令移动到 /var/www/html 目录下，操作命令如下。

```
[root@master opt]# tar -zxvf ambari-2.6.2.2-centos7.tar.gz -C /var/www/html/
[root@master opt]# tar -zxvf HDP-2.6.5.0-centos7-rpm.tar.gz -C /var/www/html/
[root@master opt]# tar -zxvf HDP-GPL-2.6.5.0-centos7-gpl.tar.gz -C /var/www/html/
[root@master opt]# mv HDP-UTILS /var/www/html/
```

④ 使用上述配置的 Web 服务器作为 yum 源服务器，配置 priorities.conf 文件内容，操作命令及文件内容如下。

```
[root@master opt]# yum -y install yum-plugin-priorities
[root@master opt]# vi /etc/yum/pluginconf.d/priorities.conf
#文件内容如下
[main]
enabled=1
gpgcheck=0
```

⑤ 将准备好的 ambari.repo 和 hdp.repo 文件移动到 /etc/yum.repos.d 目录下，编辑 ambari.repo 文件，注意 baseurl 的修改值，编辑内容如下。

```
#VERSION_NUMBER=2.6.2.2-1
[ambari-2.6.2.1]
name=ambari Version - ambari-2.6.2.1
baseurl=http://172.22.0.20/ambari/centos7/2.6.2.2-1
gpgcheck=0
gpgkey=http://172.22.0.20/ambari/centos7/2.6.6.6-1/RPM-GPG-KEY/RPM-GPG-KEY-Jenkins
enabled=1
priority=1
```

⑥ 编辑 hdp.repo 文件，注意 baseurl 的修改值，编辑内容如下。

```
#VERSION_NUMBER=2.6.5.0-292
[HDP-2.6.5.0]
name=HDP Version - HDP-2.6.5.0
baseurl=http://172.22.0.20/HDP/centos7/2.6.5.0-292/
gpgcheck=0
gpgkey=http://172.22.0.20/HDP/centos7/2.6.5.0-292/RPM-GPG-KEY/RPM-GPG-KEY-Jenkins
enabled=1
priority=1

[HDP-UTILS-1.1.0.22]
name=HDP-UTILS Version - HDP-UTILS-1.1.0.22
baseurl=http://172.22.0.20/HDP-UTILS/centos7/1.1.0.21/
gpgcheck=0
gpgkey=http://172.22.0.20/HDP-UTILS/centos7/1.1.0.21/RPM-GPG-KEY/RPM-GPG-KEY-Jenkins
enabled=1
priority=1

[HDP-GPL-2.6.5.0]
name=HDP-GPL Version - HDP-GPL-2.6.5.0
baseurl=http://172.22.0.20/HDP-GPL/centos7/2.6.5.0-292/
gpgcheck=0
gpgkey=http://172.22.0.20/HDP-GPL/centos7/2.6.5.0-292/RPM-GPG-KEY/RPM-GPG-KEY-Jenkins
enabled=1
priority=1
```

（3）配置 MySQL 数据库

Ambari Server 需要一个数据库存储元数据，默认使用 PostgreSQL 数据库。默认的用户名和密码分别是 ambari 和 bigdata。但是一般情况下，还需要安装 Hive 和 Ranger，考虑到上述需求，在此选用 MySQL 作为第三方数据库。本任务中将使用 RMP 包安装 MySQL 数据库。

① 禁用 SELinux 服务。如果不禁用 SELinux 服务，则后面启动的 MySQL 服务将会被阻止。另外，不要自行卸载 CentOS 自带的 MariaDB 数据库，安装过程中 MySQL 会自动替换 MariaDB 数据库，自行卸载 MariaDB 数据库可能会带来一些操作问题。

② 解压 mysql-5.7.27-1.el7.x86_64.rpm-bundle.tar。解压 MySQL 安装包，并在解压目录下使用"yum"命令安装 MySQL 数据库，操作命令如下。

```
[root@master opt]# tar -zxvf mysql-5.7.27-1.el7.x86_64.rpm-bundle.tar
[root@master opt]# yum install mysql-community-{server,client,common,libs}-*
```

③ 完成安装后，编辑/etc/my.cnf 文件，在"[mysqld]"配置项中添加 skip-grant-tables，如图 7-2 所示。

```
# For advice on how to change settings please see
# http://dev.mysql.com/doc/refman/5.7/en/server-configuration-defaults.html

[mysqld]
skip-grant-tables
#
# Remove leading # and set to the amount of RAM for the most important data
# cache in MySQL. Start at 70% of total RAM for dedicated server, else 10%.
# innodb_buffer_pool_size = 128M

# Remove leading # to turn on a very important data integrity option: logging
# changes to the binary log between backups.
# log_bin

# Remove leading # to set options mainly useful for reporting servers.
# The server defaults are faster for transactions and fast SELECTs.
# Adjust sizes as needed, experiment to find the optimal values.
# join_buffer_size = 128M
# sort_buffer_size = 2M
# read_rnd_buffer_size = 2M
datadir=/var/lib/mysql
socket=/var/lib/mysql/mysql.sock

# Disabling symbolic-links is recommended to prevent assorted security risks
symbolic-links=0

log-error=/var/log/mysqld.log
pid-file=/var/run/mysqld/mysqld.pid
```

图 7-2　编辑 /etc/my.cnf 文件

④ 对 MySQL 数据库进行初始化。执行 "mysqld --defaults-file=/etc/my.cnf --initialize-insecure --user=mysql" 命令，此后执行 "service mysqld restart" 命令，重启 MySQL 服务，操作命令如下。

```
[root@master opt]# mysqld --defaults-file=/etc/my.cnf --initialize-insecure
--user=mysql
[root@master opt]# service mysqld restart
```

⑤ 设置 MySQL 权限，操作命令如下。

```
[root@master ~]# mysql -u root -p
#没有密码，直接按 "Enter" 键进入 MySQL。进入后，通过执行 "set global read_only=0;" 命令，
#关闭数据库的只读属性
mysql> set global read_only=0;
mysql> flush privileges;
#执行 "set global read_only=1;" 命令，设置数据库为只读
mysql>set global read_only=1;
mysql>flush privileges;
mysql>exit;
```

⑥ 将 MySQL 的驱动 JAR 包移动到 /usr/share/java 目录下，如果该目录下没有此目录，则需要先自行创建该目录，操作命令如下。

```
[root@master opt]# mkdir -p /usr/share/java
[root@master opt]# mv mysql-connector-java-5.1.40.jar /usr/share/java/
```

（4）配置 Ambari Server

① 安装 Ambari Server，操作命令如下。

```
[root@master ~]# yum -y install ambari-server
```

编辑配置文件 ambari.properties，操作命令如下。

```
[root@master ~]# vi /etc/ambari-server/conf/ambari.properties
```

为配置 Ambari Server 与 MySQL 的连接，需要在配置文件的末尾添加如下内容。

```
server.jdbc.driver.path=/usr/share/java/mysql-connector-java-5.1.40.jar
```

② 为 Ambari 创建 MySQL 数据库和用户，操作命令如下。

```
mysql> create database ambari default character set='utf8';
mysql> CREATE USER 'ambaridba'@'localhost' IDENTIFIED BY '123456';
```

如果执行"create user 'ambaridba'@'localhost' identified by '123456';"命令时报错，则可先执行"flush privileges;"命令，再执行该命令。

```
mysql> flush privileges;
mysql> create user 'ambaridba'@'%' identified by '123456';
```

进入 MySQL，使用新建的 ambari 数据库，执行"source /var/lib/ambari-server/resources/Ambari-DDL-MySQL-CREATE.sql;"命令。

```
mysql> use ambari;
mysql> source /var/lib/ambari-server/resources/Ambari-DDL-MySQL-CREATE.sql;
#重启 MySQL 服务
[root@master ~]# service mysqld restart
```

③ 安装 Ambari Server，操作命令如下。

```
[root@master ~]# ambari-server setup
```

需要注意的是，在所有需要选择（[y/n]）的时候，都选择"y"，在提示"Checking JDK…"时选择"3"，并输入 JDK 的安装路径。安装 Ambari Server 的参数配置如图 7-3 所示。

```
[root@master ~]# ambari-server setup
Using python  /usr/bin/python
Setup ambari-server
Checking SELinux...
WARNING: Could not run /usr/sbin/sestatus: OK
Customize user account for ambari-server daemon [y/n] (n)? y
Enter user account for ambari-server daemon (root):root
Adjusting ambari-server permissions and ownership...
Checking firewall status...
Checking JDK...
[1] Oracle JDK 1.8 + Java Cryptography Extension (JCE) Policy Files 8
[2] Oracle JDK 1.7 + Java Cryptography Extension (JCE) Policy Files 7
[3] Custom JDK
==============================================================================
Enter choice (1): 3
WARNING: JDK must be installed on all hosts and JAVA_HOME must be valid on all hosts.
WARNING: JCE Policy files are required for configuring Kerberos security. If you plan
to use Kerberos,please make sure JCE Unlimited Strength Jurisdiction Policy
Files are valid on all hosts.
Path to JAVA_HOME: /usr/lib/jdk1.8.0
Validating JDK on Ambari Server...done.
Checking GPL software agreement...
GPL License for LZO: https://www.gnu.org/licenses/old-licenses/gpl-2.0.en.html
Enable Ambari Server to download and install GPL Licensed LZO packages [y/n] (n)? y
Completing setup...
Configuring database...
Enter advanced database configuration [y/n] (n)? y
Configuring database...
```

图 7-3　安装 Ambari Server 的参数配置

项目 7 Ambari 搭建与集群管理

此外,在系统提示"Choose one of the following options:"时选择"3",在"Hostname(localhost):"提示项中输入 master 容器的 IP 地址,在"Port(3306):"和"Database name(ambari):"提示项中选择默认值(直接按"Enter"键即可),在"Username(ambari):"提示项中输入"ambaridba",在"Enter Database Password(bigdata):"提示项中输入密码"123456",在"Re-enter password:"提示项中再次输入密码"123456"进行确认。配置完成后,如果可以看到返回信息中包含"successfully",则表示数据库配置成功,如图 7-4 所示。

```
Choose one of the following options:
[1] - PostgreSQL (Embedded)
[2] - Oracle
[3] - MySQL / MariaDB
[4] - PostgreSQL
[5] - Microsoft SQL Server (Tech Preview)
[6] - SQL Anywhere
[7] - BDB
================================================================
Enter choice (1): 3
Hostname (localhost): 172.22.0.20
Port (3306):
Database name (ambari):
Username (ambari): ambaridba
Enter Database Password (bigdata):
Re-enter password:
Configuring ambari database...
Configuring remote database connection properties...
WARNING: Before starting Ambari Server, you must run the following DDL
against the database to create the schema: /var/lib/ambari-server/resources/Ambari-DDL-MySQL-CREATE.sql
Proceed with configuring remote database connection properties [y/n] (y)? y
Extracting system views...
ambari-admin-2.6.2.2.1.jar
.........
Adjusting ambari-server permissions and ownership...
Ambari Server 'setup' completed successfully.
```

图 7-4 数据库配置成功

④ 执行"ambari-server start"命令,启动 ambari-server 服务,操作命令及结果如下。

```
[root@master ~]# ambari-server start
Using python  /usr/bin/python
Starting ambari-server
Ambari Server running with administrator privileges.
Organizing resource files at /var/lib/ambari-server/resources...
Ambari database consistency check started...
Server PID at: /var/run/ambari-server/ambari-server.pid
Server out at: /var/log/ambari-server/ambari-server.out
Server log at: /var/log/ambari-server/ambari-server.log
Waiting for server start....................
Server started listening on 8080

DB configs consistency check: no errors and warnings were found.
Ambari Server 'start' completed successfully.
```

访问"http://宿主机 IP 地址:8080",进入登录界面,如图 7-5 所示。

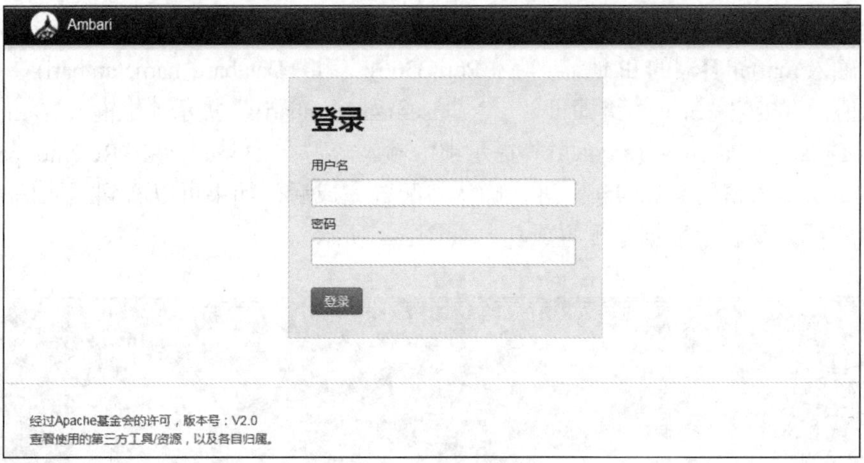

图 7-5　登录界面

⑤ 执行 "ambari-server stop" 命令，停止 ambari-server 服务，执行 "ambari-server status" 命令，查看 ambari-server 服务的状态，操作命令及结果如下。

```
[root@master ~]# ambari-server stop
[root@master ~]# ambari-server status
Using python  /usr/bin/python
Ambari-server status
Ambari Server running
Found Ambari Server PID: 258 at: /var/run/ambari-server/ambari-server.pid
```

2. 部署管理 Hadoop 集群

① 在图 7-5 所示界面中输入正确的用户名和密码（默认都是 "admin"），单击 "登录" 按钮，进入 Ambari 管理界面，如图 7-6 所示。在此界面中即可启动安装向导、创建集群和安装服务。

图 7-6　Ambari 管理界面

项目 7 Ambari 搭建与集群管理

② 单击图 7-6 所示的 Ambari 管理界面中的"启动安装向导"按钮,开始部署 Ambari。如图 7-7 所示,Ambari 安装向导提示为集群命名,此处在"Name your cluster learn more"文本框中将名称设置为"hdpCluster",然后单击"Next"按钮。

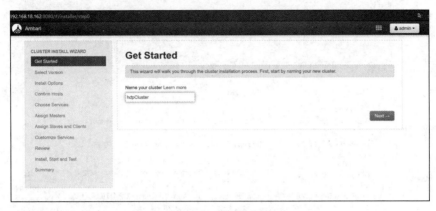

图 7-7 为集群命名

③ 进入"Select Version"界面,选择版本,如图 7-8 所示。

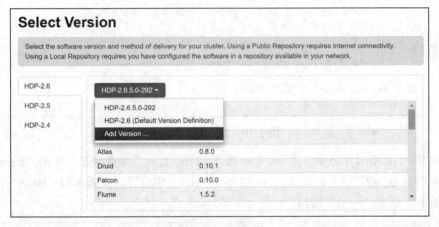

图 7-8 选择版本

④ 在图 7-8 所示界面中选择"Add Version…"选项,弹出"Add Version"对话框,勾选"Version Definition File URL"复选框,在下方的文本框中输入标示版本文件的 URL,添加版本,如图 7-9 所示。

图 7-9 添加版本

⑤ 在图 7-9 所示界面中单击"Read Version Info"按钮，进入默认设置界面，如图 7-10 所示。

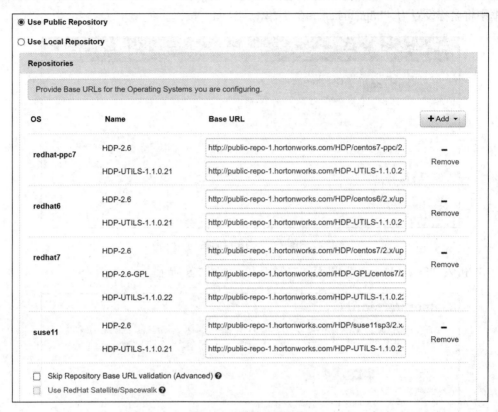

图 7-10 默认设置界面

⑥ 在图 7-10 所示界面中，选中"Use Local Repository"单选按钮，单击"Remove"按钮移除除了"redhat7"以外的版本。在"Repositories"中设置对应版本的"Base URL"，如图 7-11 所示，单击"Next"按钮。

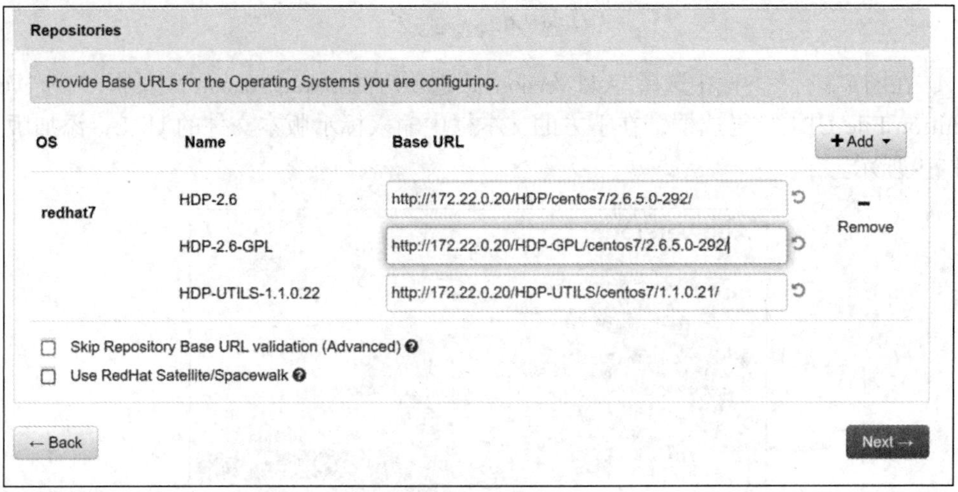

图 7-11 设置对应版本的"Base URL"

项目 7　Ambari 搭建与集群管理

⑦ 在进入的"Install Options"界面中设置"Target Hosts",在其下方的文本框中输入"master",同时设置"Host Registration Information",在其下方的文本框中粘贴 master 节点的 id_rsa 文件的内容,设置"SSH User Account"为"root",设置"SSH Port Number"为"22",完成安装选项设置,如图 7-12 所示。

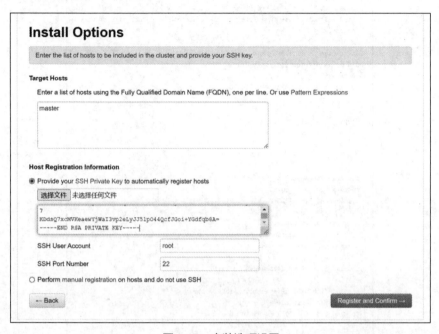

图 7-12　安装选项设置

⑧ 在图 7-12 所示界面中单击"Register and Confirm"按钮,进入"Confirm Hosts"界面,等待认证结果。在该界面中经常会出现主机认证失败的情况,即"Status"项的内容为"Failed",此时可以单击"Failed"链接,打开 master 节点的注册日志,查看失败原因,如图 7-13 和图 7-14 所示。

图 7-13　"Status"项的内容为"Failed"

197

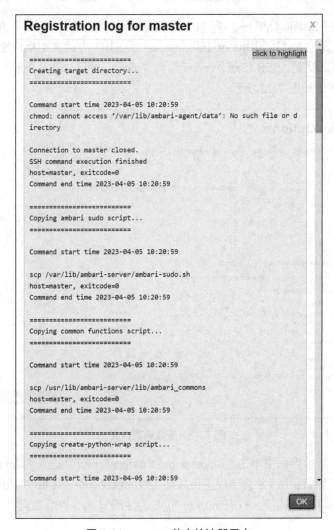

图 7-14　master 节点的注册日志

下面给出应对失败的解决方案。

修改权限，操作命令如下。

```
[root@master centos7]# chmod o=r /var/lib/ambari-agent/data
```

编辑 ambari-agent.ini 文件，操作命令及编辑文件内容如下。

```
[root@master centos7]# vi /etc/ambari-agent/conf/ambari-agent.ini
#在文件末尾添加如下配置
[security]
force_https_protocol=PROTOCOL_TLSv1_2
```

更新 OpenSSL 的版本，操作命令如下。

```
[root@master centos7]# yum install openssl
```

单击图 7-13 所示界面中的"Retry Failed"按钮，直到"Status"项的内容为"Success"，如图 7-15 所示。需要注意的是，进行认证操作的所有主机必须全部检查通过，才能保证后续操作的顺利实现。

项目 7　Ambari 搭建与集群管理

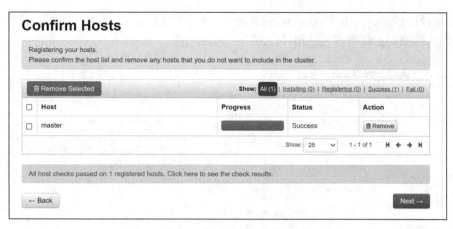

图 7-15　"Status"项的内容为"Success"

⑨ 在全部检查通过后，Ambari Agent 安装成功。单击图 7-15 所示界面中的"Next"按钮，在进入的界面中选择要安装的 Hadoop 组件。这里选择安装"HDFS""ZooKeeper"和"Ambari Metrics"3 个组件，如图 7-16 所示。

图 7-16　选择要安装的 Hadoop 组件

⑩ 选择组件后，单击"Next"按钮，进入"Assign Masters"界面，如图7-17所示。根据负载均衡的原则，将要安装的组件指派给master节点。

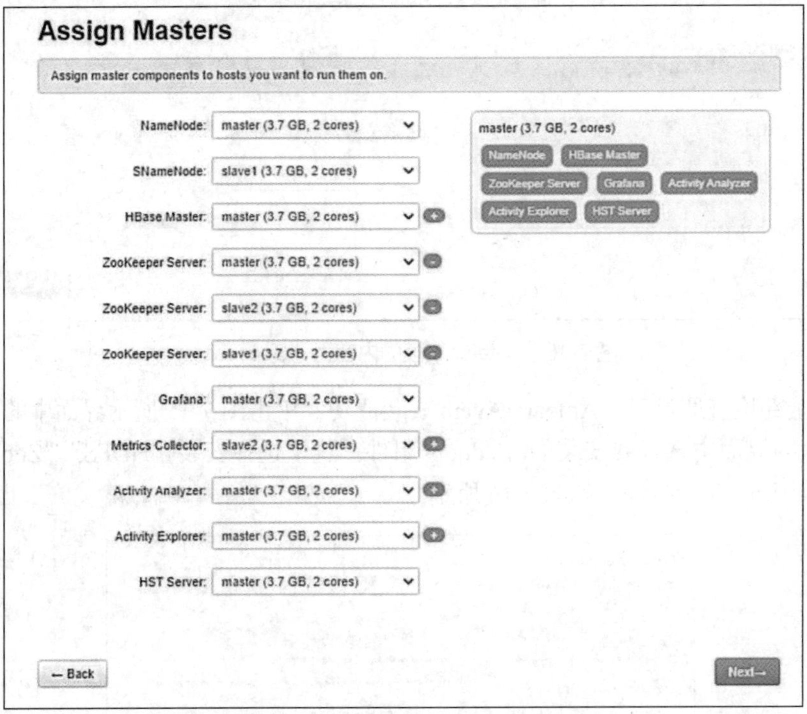

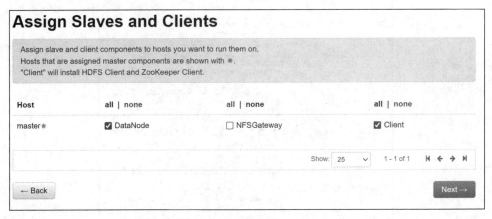

图7-17 "Assign Masters"界面

⑪ 单击图7-17所示界面中的"Next"按钮，进入"Assign Slaves and Clients"界面，如图7-18所示，根据需要指派在节点上安装的组件，这里选择安装"DataNode"和"Client"。选择完成后，单击"Next"按钮。

图7-18 "Assign Slaves and Clients"界面

⑫ 在进入的"Customize Services"界面中，查看各项参数的配置，主要关注内存配置，根据机器总内存大小来进行分配。这些参数在集群创建好后都是可以调整的，即定制服务，如图7-19所示。在图7-19中可以看到有2处提示，该提示表明需要进行进一步的设置操作。

项目 7　Ambari 搭建与集群管理

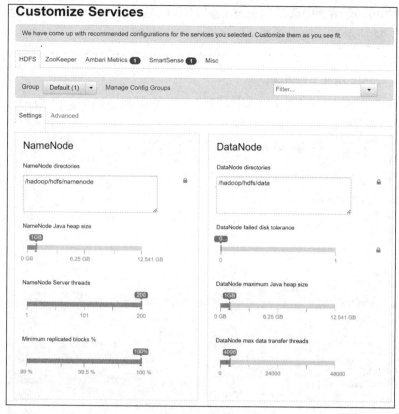

图 7-19　定制服务

⑬ 在图 7-19 所示界面中选择"Ambari Metrics"选项卡，在"Grafana Admin Password"右侧的两个文本框中分别输入默认密码"admin"，设置 Ambari Metrics 服务，如图 7-20 所示。

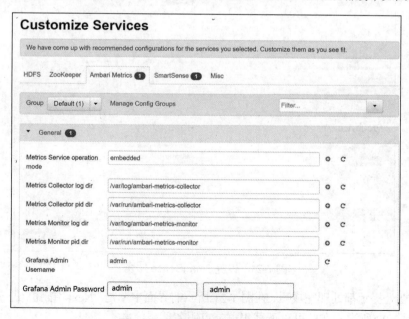

图 7-20　设置 Ambari Metrics 服务

201

⑭ 在图 7-19 所示界面中选择 "SmartSense" 选项卡，输入默认密码 "admin"，设置 SmartSense 服务，如图 7-21 所示。

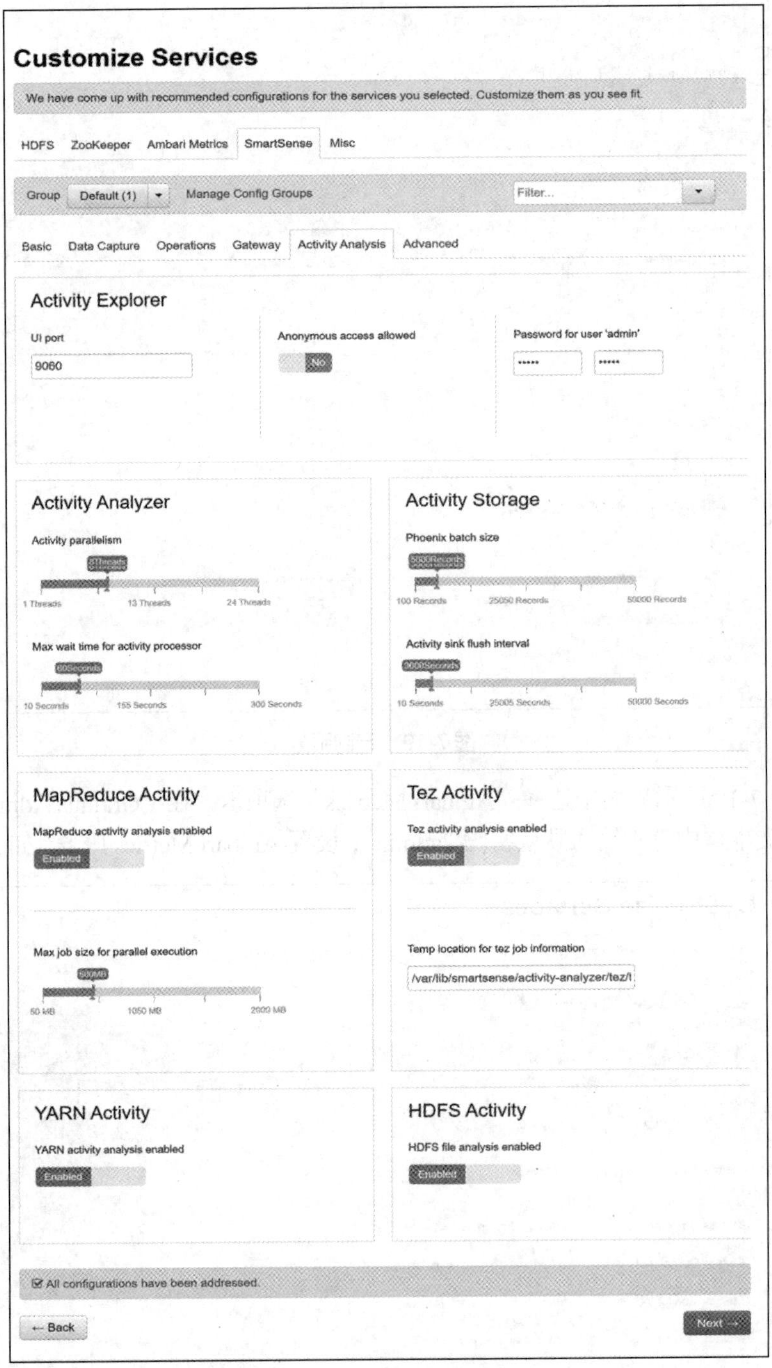

图 7-21 设置 SmartSense 服务

⑮ 在 2 处提示都处理完成后，在图 7-21 所示的界面中单击 "Next" 按钮，进入 "Review" 界面，如图 7-22 所示，在此可以下载集群节点服务部署信息。

项目 7　Ambari 搭建与集群管理

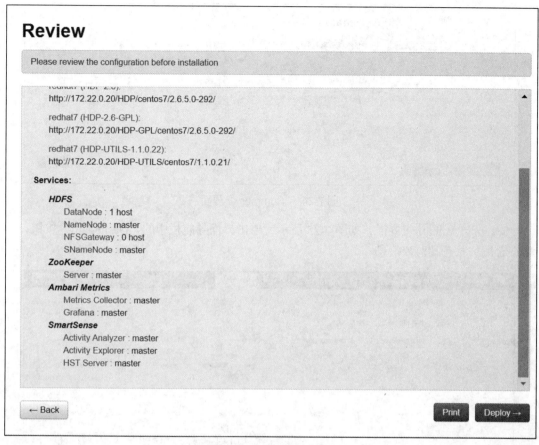

图 7-22　"Review"界面

⑯ 再次确认软件安装正确后，在图 7-22 所示界面中单击"Deploy"按钮，开始全自动部署安装，部署操作完成的服务部署界面如图 7-23 所示。

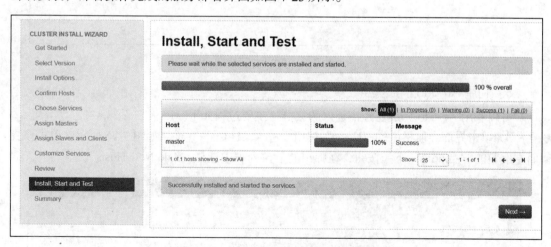

图 7-23　部署操作完成的服务部署界面

⑰ 在图 7-23 所示界面中单击"Next"按钮，进入"Summary"界面，如图 7-24 所示，查看软件安装进程的情况，并单击"Complete"按钮，完成软件的安装和部署操作。

203

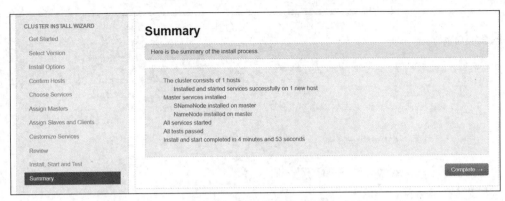

图 7-24 "Summary" 界面

⑱ 进入 Ambari 主界面,如图 7-25 所示,单击界面导航栏中的"Dashboard"按钮,可以查看集群状态和监控信息。

图 7-25 Ambari 主界面

3. 利用 Ambari 扩展集群

可以利用 Ambari 扩展已经搭建好的集群,主要过程如下。

① 单击界面导航栏中的"Hosts"按钮,进入"Hosts"界面,如图 7-26 所示,单击该界面左上角的"Actions"下拉按钮,在弹出的下拉列表中选择"Add New Hosts"选项。

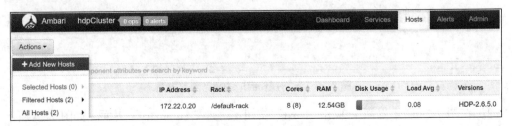

图 7-26 "Hosts" 界面

项目 7　Ambari 搭建与集群管理

② 进入"Add Host Wizard"界面，如图 7-27 所示，在"Install Options"选项卡右侧的两个文本框中分别输入需要新增的机器名（包含完整域名）和 Ambari Service 机器上生成的私钥。

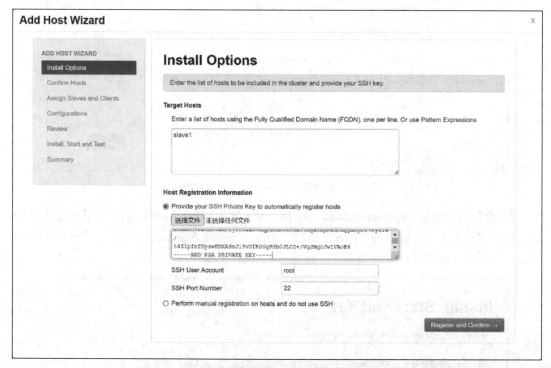

图 7-27　"Add Host Wizard"界面

③ 在图 7-27 所示界面中单击"Register and Confirm"按钮，进入"Confirm Hosts"界面，如图 7-28 所示，若"Status"项的内容显示为"Failed"，则重复节点的修改操作，并重新部署，直到"Status"项的内容显示为"Success"。

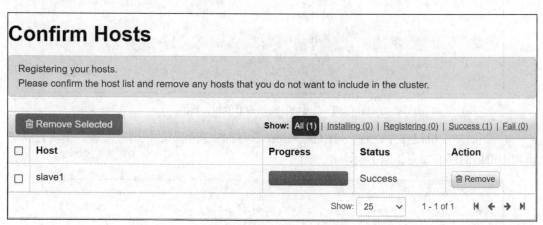

图 7-28　"Confirm Hosts"界面

④ 选择要给 slave1 节点安装的组件，如图 7-29 所示，这里选择安装"Client"，操作完成后，单击"Next"按钮。

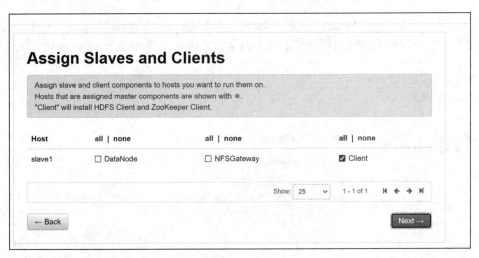

图 7-29　选择要给 slave1 节点安装的组件

⑤ 在随后进入的"Configurations"和"Review"界面中，Ambari 为用户选择了默认的配置。依次单击对应界面中的"Next"按钮，进入"Install, Start and Test"界面，如图 7-30 所示，进行服务的部署，等待部署完成后单击"Next"按钮。

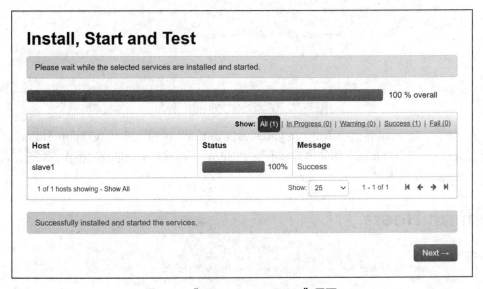

图 7-30　"Install, Start and Test"界面

⑥ 当添加主机设置完成后，可以在"Hosts"界面中看到新的机器及安装的模块，成功添加主机后的界面如图 7-31 所示。

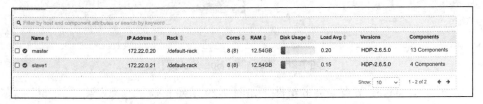

图 7-31　成功添加主机后的界面

项目 7 Ambari 搭建与集群管理

任务 7.2 使用 Ambari 管理 Hadoop 集群

任务描述

（1）学习通过 Ambari Web 管理界面对 Hadoop 服务进行管理。
（2）学习通过 Ambari Web 管理界面对集群主机进行管理。
（3）学习通过 Ambari Web 管理界面对 Hadoop 服务进程进行管理。
（4）学习通过 Ambari Web 管理界面对服务配置文件进行管理。
（5）学习使用 Ambari 完成对 Hadoop 集群的管理。

任务目标

（1）学会使用 Ambari Web 管理界面对 Hadoop 服务进行管理。
（2）学会使用 Ambari Web 管理界面对集群主机进行管理。
（3）学会使用 Ambari Web 管理界面对 Hadoop 服务进程进行管理。
（4）学会使用 Ambari Web 管理界面对服务配置文件进行管理。

知识准备

Ambari 的管理界面有助于平台管理员管理、维护和监控 Hadoop 集群，下面将介绍如何使用 Ambari Web 管理界面来进行集群管理。

在 Hadoop 集群部署完成后，打开部署 Ambari Server 主机的 8080 端口。默认的管理员用户名为 admin，密码为 admin。登录后进入的是 Ambari 管理 Hadoop 集群的主界面，该界面形象地展示了集群服务的运行状态、资源使用状况、配置参数及错误告警等信息。

任务实施

集群管理包含服务管理、主机管理、进程管理和配置管理。

微课 7-2 使用 Ambari 管理 Hadoop 集群

1. 服务管理

① 在 Ambari 主界面左侧的服务列表中，可以选择任意一个想要操作的服务。以 HDFS 为例，在左侧服务列表中选择"HDFS"选项后，可在界面右侧查看该集群服务的相关信息，如图 7-32 所示。

② 选择"Summary（概要）"选项卡，可以看到 HDFS 运行的进程信息，包括运行状态、资源使用情况及监控信息。单击界面顶部导航栏中的"Services（服务）"按钮，再单击导航栏下方的"Service Actions"（服务操作）下拉按钮，在弹出的下拉列表中有"Start""Stop""Restart"等服务控制选项，分别表示启动、停止、重启集群中所有该服务的进程，通过这些选项可以对该服务进行管理。

③ Hadoop 集群部署完成后，并不知道这个集群是否可用。此时可以借助"Run a service check（运行服务检查）"选项来查看集群服务是否正常运行。选择此选项后，会在弹出的对话框中显示服务检查情况，如图 7-33 所示。

图 7-32　查看集群服务的相关信息

图 7-33　服务检查情况

其实,这里就是通过向 HDFS 中的/tmp 目录上传一个临时文件来检查系统运行是否正常。当进度条执行完毕后,显示为绿色代表服务运行正常,显示为红色代表服务运行失败,显示为黄色代表出现告警信息。

④ 在"Service Actions"下拉列表中选择"Restart"选项,可以重启集群中所有 HDFS 服务的进程。进入 HDFS 服务重启界面后,可以查看主机进程的操作进度和运行日志,如图 7-34 所示。

图 7-34　查看主机进程的操作进度和运行日志

在"Service Actions"下拉列表中有"Turn On Maintenance Mode（打开维护模式）"选项,该选项用于在用户调试或者维护过程中抑制不必要的告警信息的产生,以及避免因批量操作而对服务产生影响（启动所有服务、停止所有服务、重启所有服务等）。维护模式中设置有不同的级别,分别是服务级别、主机级别及进程级别。这 3 种级别之间存在着覆盖关系。例如,由于 HDFS 部署在多台主机中,因此当它的维护模式功能启用后,HDFS 便不会产生任何新的告警信息。当用户重启集群的所有服务时,这些服务不会受到批量操作的影响。当用户重启一个机器的所有进程时,该服务的进程也不会被批量操作影响。

⑤ 在 Ambari 主界面左侧的服务列表的最下方有一个"Actions（动作）"下拉按钮,单击该下拉按钮,可以弹出对服务进行操作的下拉列表,其中包含"增加服务（Additional services）""启动所有服务（Start all services）""停止所有服务（Stop all services）"等选项。

2. 主机管理

① 单击 Ambari 主界面顶部导航栏中的"Hosts（主机）"按钮,可以打开 Ambari 管理的主机列表,如图 7-35 所示。

② 单击 Ambari 主界面导航栏左下方的"Actions"下拉按钮,在弹出的下拉列表中可以选择与主机相关的操作选项,其作用与"Service Actions"类似,只是执行的范围不同。

主机管理操作如图 7-36 所示。

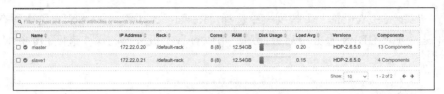

图 7-35　主机列表

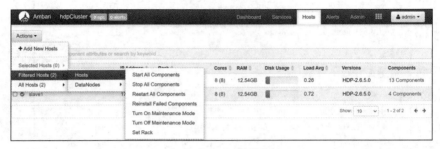

图 7-36　主机管理操作

当用户在"Actions"下拉列表中选择"Filtered Hosts（显示主机）"→"Hosts"→"Start All Components（启动所有组件）"选项时，Ambari 会启动主机中的所有服务。

当用户在"Actions"下拉列表中选择"All Hosts（所有主机）"→"DataNodes"→"Stop All Components（停止所有组件）"选项时，Ambari 会关闭所有机器关于 DataNode 的进程。

③ 当集群不能提供生产环境所需的足够资源时，可以通过"Actions"下拉列表中的"Add New Hosts（添加新的主机）"进项来扩展集群。新的主机节点在加入集群之前，需要先完成任务 6.1 中基本环境的配置及 Ambari Agent 服务的安装配置操作。

④ 进入其中一台主机（如 master）的界面查看主机信息，可以查看该主机中所有进程的运行状态、主机资源使用情况、主机的 IP 地址、资源栈等信息，如图 7-37 所示。

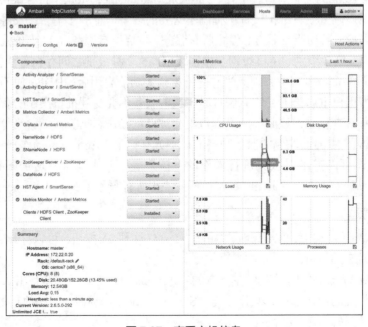

图 7-37　查看主机信息

项目 7　Ambari 搭建与集群管理

⑤ 在 Ambari 主界面导航栏右下方单击"Host Actions（主机动作）"下拉按钮，在弹出的下拉列表中可以选择"Turn On Maintenance Mode（打开维护模式）"选项，对于主机级别的维护模式来说，打开维护模式表示打开该主机所有进程的维护模式。如果该主机已经存在告警信息，那么一旦维护模式被打开，告警信息就会被屏蔽，并且不会产生新告警信息，所有的批量操作都会忽略该机器。

3. 进程管理

每个服务都由相应的进程组成，如 HDFS 服务包含 NameNode、SecondaryNameNode、DateNode 等进程。每台主机中都安装了相应的服务进程，如 master 节点中包含 HDFS 服务的 NameNode 进程，slave1 节点中包含 SecondaryNameNode、DateNode 进程。

进入 master 节点，找到需要进行管理的进程，如 NameNode，该进程后面的"Started"下拉按钮表示该进程正在运行中，单击该按钮可以改变进程的运行状态，如弹出的下拉列表中的"Restart""Stop""Turn On Maintenance Mode""Delete"分别表示重启该进程、停止该进程、打开该进程的维护模式和删除该进程。进程管理操作如图 7-38 所示。

其中，打开进程级别的维护模式后会产生以下两种影响。

（1）该进程不再受批量操作的控制。

（2）抑制该进程告警信息的产生。

例如，打开主机 master 节点的 DataNode 的维护模式，那么当用户在 Activity Explorer 对应的"Started"下拉列表中选择"Stop"选项时，将会停止所有 HDFS 服务，但该主机的 DataNode 不会被关闭，这是因为停止 HDFS 服务的批量操作后，会直接忽略 master 节点中的 DataNode。

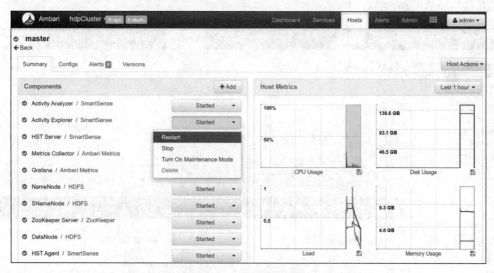

图 7-38　进程管理操作

4. 配置管理

Ambari Web 管理界面可以修改配置文件，并将其应用到集群的每台主机中，在集群的主机数量非常多时，这种方式十分方便。

例如，需要将集群 HDFS 的 Block 复制（Block Replication）因子修改为 2，在手动部署集群的情况下，要修改每一台主机的 hdfs-site.xml 配置文件。如果一个集群中有几十台或者几百台主机，则工作量是非常大的。而 Ambari Web 管理界面可以很好地应对这种情况，由集群中的 Ambari Server 向每台主机的 Ambari Agent 发送相关的心跳信号，从而更新每台主机中的配置文件。具体操作步骤如下。

① 在 Ambari 主界面的导航栏中选择"Services"→"HDFS"选项，在其服务列表右侧选择"Configs"→"Advanced"→"General"→"Block Replication"选项，修改 Block 复制因子，将"Block replication"修改为 2，如图 7-39 所示，单击"Save"按钮。

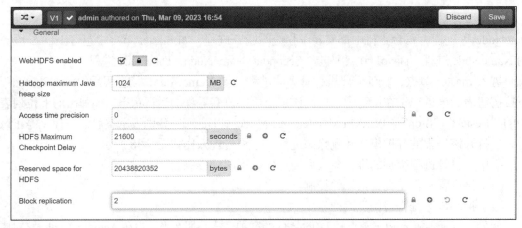

图 7-39　修改 Block 复制因子

② 保存成功后，可以查看相应的版本信息，如图 7-40 所示，单击"Restart"按钮，重启所有标记重启的组件。

图 7-40　查看相应的版本信息

③ 在 Ambari Web 管理界面中，可以查看某一台主机的服务配置信息。单击 Ambari 主界面导航栏中的"Hosts"按钮，在进入的界面中选择相应的主机，选择"Configs"选项卡，便可以查看主机的服务配置信息。为了保证整个集群配置信息的统一，这里要禁止单独修改某一台主机的配置文件。

项目 7　Ambari 搭建与集群管理

项目小结

本项目介绍了 Ambari 的功能、NTP 的安装与配置、本地仓库的安装与配置、MySQL 数据库配置、Ambari Server 的安装与配置等，并介绍了使用 Ambari 进行 Hadoop 组件的安装、配置和运维等，完成了利用 Ambari 扩展集群的操作。学习完本项目，读者能够使用 Ambari 安装和管理 Hadoop 集群。

课后练习

1. 请简述 Ambari 的作用。
2. Ambari 中的告警类型有哪些?
3. 启动、重启、关闭 Ambari Server 和 Ambari Agent 的命令分别是什么?

项目 ❽ Hadoop 平台应用综合案例

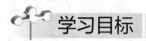

【知识目标】
熟悉 HDFS、Hive、MySQL、HBase 的数据互导。
了解 Hive 与传统关系数据库的区别。
熟悉 Hive 的技术优势。

【素质目标】
具有严谨细致的工作态度和工作作风。
具有良好的团队协作意识和业务沟通能力。
具有良好的表达能力和文档查阅能力。
具有规范的编程意识和较好的数据洞察力。

【技能目标】
学会 HDFS、Hive、MySQL、HBase 的数据互导操作。
学会使用 Hive 进行简单的数据分析操作。
学会使用 Flume、Kafka、Flink 进行简单的模拟流数据处理。

项目描述

在 Hadoop 平台应用中，为了进行数据处理，往往需要在各个组件间进行数据的导入与导出操作，并利用 Hive 进行简单的数据分析。

本项目主要完成以下任务。
（1）将本地数据集上传到 Hive 中。
（2）使用 Hive 进行简单的数据分析。
（3）实现 Hive、MySQL、HBase 数据的互导。
（4）使用 Flume、Kafka、Flink 实现简单的模拟流数据处理。

任务 8.1　本地数据集上传到 Hive 中

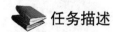

将本地数据集上传到 Hive 中。

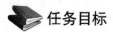

学会将本地数据集上传到 Hive 中。

项目 8　Hadoop 平台应用综合案例

任务实施

本任务将 test.txt 中的数据上传到 Hive 中。为了完成这个操作，需要先把 test.txt 中的数据上传到 HDFS 中，再在 Hive 中创建一个外部表，完成数据的导入。

微课 8-1　本地数据集上传到 Hive 中

1. 启动 HDFS

① HDFS 是 Hadoop 的核心组件，因此，要想使用 HDFS，必须先安装 Hadoop。在前面的任务中已经安装了 Hadoop，所以接下来打开一个终端，执行 "start-all.sh" 命令，启动 Hadoop 服务即可，操作命令如下。

```
[root@master1 ~]# start-all.sh
```

② 执行 "jps" 命令，查看当前运行的进程，操作命令及结果如下。

```
[root@master1 ~]# jps
673 ResourceManager
503 SecondaryNameNode
935 Jps
298 NameNode
[root@slave1 ~]# jps
256 NodeManager
388 Jps
140 DataNode
[root@slave2 ~]# jps
243 NodeManager
375 Jps
127 DataNode
```

2. 将本地文件上传到 HDFS 中

① 将本地文件 test.txt 上传到 HDFS 中，并将其存储在 HDFS 的/bigdatacase/dataset 目录中。在 HDFS 的根目录中创建一个新的目录 bigdatacase，并在其中创建一个子目录 dataset，操作命令及结果如下。

```
[root@master1 ~]# hadoop dfs -mkdir -p /bigdatacase/dataset
DEPRECATED: Use of this script to execute hdfs command is deprecated.
Instead use the hdfs command for it.
```

② 执行 "hadoop dfs -put /opt/test.txt /bigdatacase/dataset" 命令，将 test.txt 文件上传到 HDFS 的/bigdatacase/dataset 目录中，操作命令及结果如下。

```
[root@master1 ~]# hadoop dfs -put /opt/test.txt /bigdatacase/dataset
DEPRECATED: Use of this script to execute hdfs command is deprecated.
Instead use the hdfs command for it.
```

③ 执行 "hadoop dfs -cat /bigdatacase/dataset/test.txt | head -10" 命令，查看 HDFS 中的 test.txt 的前 10 条记录，操作命令及结果如下。

```
[root@master1 ~]# hadoop dfs -cat /bigdatacase/dataset/test.txt | head -10
```

```
DEPRECATED: Use of this script to execute hdfs command is deprecated.
Instead use the hdfs command for it.
27.19.74.143,2015/3/30 17:38,/static/image/common/faq.gif
110.52.250.126,2015/3/30 17:38,/data/cache/style_1_widthauto.css?y7a
27.19.74.143,2015/3/30 17:38,/static/image/common/hot_1.gif
27.19.74.143,2015/3/30 17:38,/static/image/common/hot_2.gif
27.19.74.143,2015/3/30 17:38,/static/image/filetype/common.gif
110.52.250.126,2015/3/30 17:38,/source/plugin/wsh_wx/img/wsh_zk.css
110.52.250.126,2015/3/30 17:38,/data/cache/style_1_forum_index.css?y7a
110.52.250.126,2015/3/30 17:38,/source/plugin/wsh_wx/img/wx_jqr.gif
27.19.74.143,2015/3/30 17:38,/static/image/common/recommend_1.gif
110.52.250.126,2015/3/30 17:38,/static/image/common/logo.png
```

3. 在 Hive 中创建数据库

（1）创建数据库和数据表

① 执行 "service mysql start" 命令，启动 MySQL 数据库，操作命令及结果如下。

```
[root@master1 ~]# service mysql start
Starting MySQL SUCCESS!
```

② Hive 是基于 Hadoop 的数据库，使用 HiveQL 编写的查询语句最终都会被 Hive 自动解析为 MapReduce 任务，并由 Hadoop 执行。因此，需要先启动 Hadoop 服务，再启动 Hive 服务。可通过执行 "./hive" 命令来启动 Hive 服务，操作命令及结果如下。

```
[root@master1 bin]# ./hive
Logging initialized using configuration in jar:file:/simple/hive1.2.1/lib/
hive-common-1.2.1.jar!/hive-log4j.properties
hive>
```

③ 启动 Hive 服务后，执行 "create database dblab;" 命令，在 Hive 中创建一个数据库 dblab，操作命令及结果如下。

```
hive> create database dblab;
OK
Time taken: 0.592 seconds
```

④ 创建外部表，操作命令及结果如下。

```
hive> create external table dblab.bigdata_user(ip string,time string,url string)
row format delimited fields terminated by ',' stored as textfile location
'/bigdatacase/dataset';
OK
Time taken: 0.042 seconds
```

（2）查询数据

① 在 Hive 命令行模式下，执行 "show create table bigdata_user;" 命令，查看表的属性，操作命令及结果如下。

```
hive> show create table bigdata_user;
OK
```

项目 8　Hadoop 平台应用综合案例

```
CREATE EXTERNAL TABLE `bigdata_user`(
 `ip` string,
 `time` string,
 `url` string)
ROW FORMAT DELIMITED
 FIELDS TERMINATED BY ','
STORED AS INPUTFORMAT
 'org.apache.hadoop.mapred.TextInputFormat'
OUTPUTFORMAT
 'org.apache.hadoop.hive.ql.io.HiveIgnoreKeyTextOutputFormat'
LOCATION
 'hdfs://master1:9000/bigdatacase/dataset'
TBLPROPERTIES (
 'COLUMN_STATS_ACCURATE'='false',
 'numFiles'='0',
 'numRows'='-1',
 'rawDataSize'='-1',
 'totalSize'='0',
 'transient_lastDdlTime'='1677831958')
Time taken: 0.033 seconds, Fetched: 19 row(s)
```

② 执行 "desc bigdata_user;" 命令，查看表的简单结构，操作命令及结果如下。

```
hive> desc bigdata_user;
OK
ip                      string
time                    string
url                     string
Time taken: 0.049 seconds, Fetched: 3 row(s)
```

③ 执行 "select * from bigdata_user limit 10;" 命令，查看表的前 10 条记录，操作命令及结果如下。

```
hive> select * from bigdata_user limit 10;
OK
27.19.74.143     2015/3/30 17:38  /static/image/common/faq.gif
110.52.250.126   2015/3/30 17:38  /data/cache/style_1_widthauto.css?y7a
27.19.74.143     2015/3/30 17:38  /static/image/common/hot_1.gif
27.19.74.143     2015/3/30 17:38  /static/image/common/hot_2.gif
27.19.74.143     2015/3/30 17:38  /static/image/filetype/common.gif
110.52.250.126   2015/3/30 17:38  /source/plugin/wsh_wx/img/wsh_zk.css
110.52.250.126   2015/3/30 17:38  /data/cache/style_1_forum_index.css?y7a
110.52.250.126   2015/3/30 17:38  /source/plugin/wsh_wx/img/wx_jqr.gif
27.19.74.143     2015/3/30 17:38  /static/image/common/recommend_1.gif
110.52.250.126   2015/3/30 17:38  /static/image/common/logo.png
Time taken: 0.046 seconds, Fetched: 10 row(s)
```

任务 8.2 使用 Hive 进行简单的数据分析

任务描述

使用 Hive 进行简单的数据分析。

任务目标

学会使用 Hive 进行简单的数据分析。

任务实施

微课8-2 使用Hive进行简单的数据分析

1. 简单查询分析

执行 "select ip from bigdata_user limit 10;" 命令，查询表的前 10 条记录中的 IP 地址，操作命令及结果如下。

```
hive> select ip from bigdata_user limit 10;
OK
27.19.74.143
110.52.250.126
27.19.74.143
27.19.74.143
27.19.74.143
110.52.250.126
110.52.250.126
110.52.250.126
27.19.74.143
110.52.250.126
Time taken: 0.064 seconds, Fetched: 10 row(s)
```

2. 查询前 20 条记录中的 IP 地址和时间

执行 "select ip,time from bigdata_user limit 20;" 命令，查询表的前 20 条记录中的 IP 地址和时间，操作命令及结果如下。

```
hive> select ip,time from bigdata_user limit 20;
OK
27.19.74.143      2015/3/30 17:38
110.52.250.126    2015/3/30 17:38
27.19.74.143      2015/3/30 17:38
27.19.74.143      2015/3/30 17:38
27.19.74.143      2015/3/30 17:38
110.52.250.126    2015/3/30 17:38
110.52.250.126    2015/3/30 17:38
```

项目 8　Hadoop 平台应用综合案例

```
110.52.250.126    2015/3/30  17:38
27.19.74.143      2015/3/30  17:38
110.52.250.126    2015/3/30  17:38
27.19.74.143      2015/3/30  17:38
110.52.250.126    2015/3/30  17:38
8.35.201.144      2015/3/30  17:38
27.19.74.143      2015/3/30  17:38
27.19.74.143      2015/3/30  17:38
27.19.74.143      2015/3/30  17:38
27.19.74.143      2015/3/30  17:38
8.35.201.165      2015/3/30  17:38
8.35.201.164      2015/3/30  17:38
8.35.201.163      2015/3/30  17:38
Time taken: 0.047 seconds, Fetched: 20 row(s)
```

3. 使用聚合函数 count() 统计表中的数据

执行 "select count(*) from bigdata_user;" 命令，统计表中的数据，操作命令及结果如下。

```
hive> select count(*) from bigdata_user;
Query ID = root_20230303162953_b096cca5-c29b-4504-acd0-231707387187
Total jobs = 1
Launching Job 1 out of 1
Number of reduce tasks determined at compile time: 1
In order to change the average load for a reducer (in bytes):
  set hive.exec.reducers.bytes.per.reducer=<number>
In order to limit the maximum number of reducers:
  set hive.exec.reducers.max=<number>
In order to set a constant number of reducers:
  set mapreduce.job.reduces=<number>
Starting Job = job_1677830821878_0001, Tracking URL = http://master1:8088/proxy/application_1677830821878_0001/
Kill Command = /usr/local/hadoop/bin/hadoop job  -kill job_1677830821878_0001
Hadoop job information for Stage-1: number of mappers: 1; number of reducers: 1
2023-03-03 16:29:59,890 Stage-1 map = 0%,  reduce = 0%
2023-03-03 16:30:05,256 Stage-1 map = 100%,  reduce = 0%, Cumulative CPU 1.17 sec
2023-03-03 16:30:10,396 Stage-1 map = 100%,  reduce = 100%, Cumulative CPU 2.38 sec
MapReduce Total cumulative CPU time: 2 seconds 380 msec
Ended Job = job_1677830821878_0001
MapReduce Jobs Launched:
Stage-Stage-1: Map: 1  Reduce: 1   Cumulative CPU: 2.38 sec   HDFS Read: 8691 HDFS Write: 3 SUCCESS
Total MapReduce CPU Time Spent: 2 seconds 380 msec
```

任务 8.3　Hive、MySQL、HBase 数据的互导

任务描述

Hive、MySQL、HBase 数据的互导。

任务目标

学会 Hive、MySQL、HBase 的数据互导操作。

任务实施

微课 8-3　Hive、MySQL、HBase 数据的互导

1. Hive 预操作

① 创建临时表 user_action，操作命令及结果如下。

```
hive> create external table user_action(ip string,time string,url string) row format delimited fields terminated by ',' stored as textfile;
OK
Time taken: 0.054 seconds
```

创建完成后，Hive 会自动在 HDFS 中创建对应的数据文件/user/hive/warehouse/dblab.db/user_action。

② 执行 "hadoop dfs -ls /user/hive/warehouse/dblab.db/" 命令，在 HDFS 中查看创建的 user_action 表，操作命令及结果如下。

```
[root@master1 /]# hadoop dfs -ls /user/hive/warehouse/dblab.db/
DEPRECATED: Use of this script to execute hdfs command is deprecated.
Instead use the hdfs command for it.
Found 1 items
drwxr-xr-x   - root supergroup          0 2023-03-03 16:32
```

2. 数据导入操作

① 在 Hive Shell 模式下执行 "insert overwrite table dblab.user_action select * from dblab.bigdata_user;" 命令，将 bigdata_user 表中的数据导入 user_action 表中，操作命令如下。

```
hive> insert overwrite table dblab.user_action select * from dblab.bigdata_user;
```

② 执行 "select * from user_action limit 10;" 命令，查询表的前 10 条记录，操作命令及结果如下。

```
hive> select * from user_action limit 10;
OK
27.19.74.143    2015/3/30 17:38 /static/image/common/faq.gif
110.52.250.126  2015/3/30 17:38 /data/cache/style_1_widthauto.css?y7a
27.19.74.143    2015/3/30 17:38 /static/image/common/hot_1.gif
```

项目 8　Hadoop 平台应用综合案例

```
27.19.74.143    2015/3/30 17:38 /static/image/common/hot_2.gif
27.19.74.143    2015/3/30 17:38 /static/image/filetype/common.gif
110.52.250.126  2015/3/30 17:38 /source/plugin/wsh_wx/img/wsh_zk.css
110.52.250.126  2015/3/30 17:38 /data/cache/style_1_forum_index.css?y7a
110.52.250.126  2015/3/30 17:38 /source/plugin/wsh_wx/img/wx_jqr.gif
27.19.74.143    2015/3/30 17:38 /static/image/common/recommend_1.gif
110.52.250.126  2015/3/30 17:38 /static/image/common/logo.png
Time taken: 0.03 seconds, Fetched: 10 row(s)
```

3. 使用 Sqoop 将数据从 Hive 导入 MySQL 中

① 登录 MySQL，在 dblab 数据库中创建与 Hive 对应的 user_action 表，并设置其编码格式为 UTF-8，操作命令及结果如下。

```
mysql> use dblab;
Database changed
mysql> create table user_action(
    -> ip varchar(50),
    -> time varchar(50),
    -> url varchar(255))
    -> ENGINE=InnoDB DEFAULT CHARSET=utf8;
Query OK, 0 rows affected (0.00 sec)
```

② 退出 MySQL，进入 sqoop/bin 目录，导入数据，操作命令如下。

```
[root@master1 bin]# ./sqoop export --connect jdbc:mysql://master1:3306/dblab
--username root --password 123456 --table user_action --export-dir
/user/hive/warehouse/dblab.db/user_action --input-fields-terminated-by ',';
```

③ 使用 root 用户登录 MySQL，查看已经从 Hive 导入 MySQL 中的数据，操作命令及结果如下。

```
mysql> select * from user_action limit 7;
+--------------+-----------------+----------------------------------------+
| ip           | time            | url                                    |
+--------------+-----------------+----------------------------------------+
| 8.35.201.163 | 2015/3/30 17:38 | /static/image/common/arw_r.gif         |
| 8.35.201.166 | 2015/3/30 17:38 | /static/image/common/px.png            |
| 8.35.201.144 | 2015/3/30 17:38 | /static/image/common/pmto.gif          |
| 8.35.201.161 | 2015/3/30 17:38 | /static/image/common/search.png        |
| 8.35.201.163 | 2015/3/30 17:38 | /uc_server/avatar.php?uid=57232&size=middle |
| 8.35.201.164 | 2015/3/30 17:38 | /uc_server/data/avatar/000/05/83/
35_avatar_middle.jpg |
| 8.35.201.160 | 2015/3/30 17:38 | /uc_server/data/avatar/000/01/54/
22_avatar_middle.jpg |
+--------------+-----------------+----------------------------------------+
7 rows in set (0.00 sec)
```

4. 使用 Sqoop 将数据从 MySQL 导入 HBase 中

① 启动 Hadoop 集群和 HBase 服务，操作命令如下。

```
[root@master1 bin]# start-all.sh
[root@master1 bin]# ./zkServer.sh start
[root@master1 bin]# ./start-hbase.sh
```

② 在 master1、slave1 和 slave2 节点上分别执行"jps"命令，查看集群节点的进程信息，master1 节点的操作命令及结果如下。

```
[root@master1 bin]# jps
1714 SecondaryNameNode
4437 Jps
3207 HMaster
1514 NameNode
1883 ResourceManager
3358 HRegionServer
3039 QuorumPeerMain
```

slave1 节点的操作命令及结果如下。

```
[root@slave1 bin]# jps
577 NodeManager
786 QuorumPeerMain
1811 Jps
854 HRegionServer
473 DataNode
```

slave2 节点的操作命令及结果如下。

```
[root@slave2 bin]# jps
578 NodeManager
3154 Jps
1028 QuorumPeerMain
474 DataNode
1102 HRegionServer
```

③ 进入 HBase Shell，操作命令及结果如下。

```
[root@master1 bin]# ./hbase shell
SLF4J: Class path contains multiple SLF4J bindings.
SLF4J: Found binding in [jar:file:/usr/local/hbase/lib/slf4j-log4j12-1.7.25.jar!/org/slf4j/impl/StaticLoggerBinder.class]
SLF4J: Found binding in [jar:file:/usr/local/hadoop/share/hadoop/common/lib/slf4j-log4j12-1.7.10.jar!/org/slf4j/impl/StaticLoggerBinder.class]
SLF4J: Actual binding is of type [org.slf4j.impl.Log4jLoggerFactory]
HBase Shell
Use "help" to get list of supported commands.
Use "exit" to quit this interactive shell.
Version 1.6.0, r5ec5a5b115ee36fb28903667c008218abd21b3f5, Fri Feb 14 12:00:03 PST 2020
```

```
hbase(main):001:0>
```

④ 在 HBase 中创建 user_action 表，操作命令及结果如下。

```
hbase(main):001:0> create 'user_action',{NAME =>'f1',VERSION =>5}
0 row(s) in 1.4250 seconds
```

⑤ 新建一个终端，导入数据，操作命令如下。

```
[root@master1 bin]# ./sqoop import --connect jdbc:mysql://localhost:3306/
dblab?zeroDateTimeBehavior=ROUND --username root --password 123456 --table
user_action --hbase-table user_action --column-family f1 --hbase-row-key ip -m 1
```

⑥ 再次切换到 HBase Shell 运行的终端窗口，执行“scan 'user_action'”命令，查询插入的数据，如图 8-1 所示。

```
hbase(main):001:0> scan 'user_action'
ROW                              COLUMN+CELL
 110.52.250.126                  column=f1:time, timestamp=1677990580371, value=2015/3/30 17:38
 110.52.250.126                  column=f1:url, timestamp=1677990580371, value=/static/js/logging.js?y7a
 27.19.74.143                    column=f1:time, timestamp=1677990580371, value=2015/3/30 17:38
 27.19.74.143                    column=f1:url, timestamp=1677990580371, value=/static/image/editor/editor.gif
 8.35.201.144                    column=f1:time, timestamp=1677990580371, value=2015/3/30 17:38
 8.35.201.144                    column=f1:url, timestamp=1677990580371, value=/uc_server/avatar.php?uid=29331&size=middl
                                 e
 8.35.201.160                    column=f1:time, timestamp=1677990580371, value=2015/3/30 17:38
 8.35.201.160                    column=f1:url, timestamp=1677990580371, value=/uc_server/data/avatar/000/04/12/85_avatar
                                 _middle.jpg
 8.35.201.161                    column=f1:time, timestamp=1677990580371, value=2015/3/30 17:38
 8.35.201.161                    column=f1:url, timestamp=1677990580371, value=/static/image/common/search.png
 8.35.201.163                    column=f1:time, timestamp=1677990580371, value=2015/3/30 17:38
 8.35.201.163                    column=f1:url, timestamp=1677990580371, value=/uc_server/data/avatar/000/04/87/94_avatar
                                 _middle.jpg
 8.35.201.164                    column=f1:time, timestamp=1677990580371, value=2015/3/30 17:38
 8.35.201.164                    column=f1:url, timestamp=1677990580371, value=/uc_server/avatar.php?uid=53635&size=middl
                                 e
 8.35.201.165                    column=f1:time, timestamp=1677990580371, value=2015/3/30 17:38
 8.35.201.165                    column=f1:url, timestamp=1677990580371, value=/uc_server/data/avatar/000/01/01/03_avatar
                                 _middle.jpg
 8.35.201.166                    column=f1:time, timestamp=1677990580371, value=2015/3/30 17:38
 8.35.201.166                    column=f1:url, timestamp=1677990580371, value=/static/image/common/px.png
9 row(s) in 0.2040 seconds
```

图 8-1　查询插入的数据

5. 利用 hbase-thrift 库将数据导入 HBase 中

① 使用“pip”命令安装最新版本的 hbase-thrift 库，操作命令如下。

```
[root@master1 bin]# pip install hbase-thrift
```

② 在 hbase/bin 目录下启动 thrift 相关命令，启用 9095 端口，操作命令如下。

```
[root@master1 bin]# hbase-daemon.sh start thrift --infoport 9095 -p 9090
```

③ 查看 9095 端口，操作命令及结果如下。

```
[root@master1 ~]# netstat -anp |grep 9095
tcp        0      0 0.0.0.0:9095        0.0.0.0:*          LISTEN      1802/java
```

④ 执行“jps”命令，查看进程运行情况，操作命令及结果如下。

```
[root@master1 ~]# jps
501 SecondaryNameNode
1125 HMaster
280 NameNode
1802 ThriftServer
1292 HRegionServer
942 QuorumPeerMain
1886 Jps
671 ResourceManager
```

⑤ 在 HBase 中创建 "ip_info" 表，表中包含两个列族，分别为 "ip" 和 "url"，并查看创建的表，操作命令及结果如下。

```
hbase(main):004:0> create 'ip_info','ip','url'
=> ["ip_info", "user_action"]
hbase(main):005:0> desc 'ip_info'
Table ip_info is ENABLED
ip_info
COLUMN FAMILIES DESCRIPTION
{NAME => 'ip', BLOOMFILTER => 'ROW', VERSIONS => '1', IN_MEMORY => 'false',
KEEP_DELETED_CELLS => 'FALSE', DATA_BLOCK_ENCODING => 'NONE', TTL => 'FOREVER',
COMPRESSION => 'NONE', MIN_VERSIONS => '0', BLOCKCACHE => 'true', BLOCKSIZE =>
'65536', REPLICATION_SCOPE => '0'}
{NAME => 'url', BLOOMFILTER => 'ROW', VERSIONS => '1', IN_MEMORY => 'false',
KEEP_DELETED_CELLS => 'FALSE', DATA_BLOCK_ENCODING => 'NONE', TTL => 'FOREVER',
COMPRESSION => 'NONE', MIN_VERSIONS => '0', BLOCKCACHE => 'true', BLOCKSIZE =>
'65536', REPLICATION_SCOPE => '0'}
2 row(s) in 0.0820 seconds
```

⑥ 使用 vim 新建并编辑 text.py 文件，文件内容如下。

```python
# -*- coding: utf-8 -*-
from thrift.transport import TSocket
from thrift.transport import TTransport
from thrift.protocol import TBinaryProtocol

import logging
from hbase import Hbase
from hbase.ttypes import ColumnDescriptor, Mutation
logging.basicConfig();
class HbaseClient(object):
    def __init__(self, host='localhost', port=9090):
        transport = TTransport.TBufferedTransport(TSocket.TSocket(host, port))
        protocol = TBinaryProtocol.TBinaryProtocol(transport)
        self.client = Hbase.Client(protocol)
        transport.open()

    def put(self, table, row, columns):

        self.client.mutateRow(table, row, map(lambda (k,v): Mutation(column=k,
value=v), columns.items())
)

    def scan(self, table, start_row="", columns=None):
```

```
            scanner = self.client.scannerOpen(table, start_row, columns)
            while True:
                r = self.client.scannerGet(scanner)
                if not r:
                    break
                yield dict(map(lambda (k,v): (k, v.value),r[0].columns.items()))
if __name__ == "__main__":
    client = HbaseClient("172.19.0.2", 9090)
    client.put("ip_info", "1", {"ip:":"172.0.1.100","url:":"/static/image/common/search.png"})
    client.put("ip_info", "2", {"ip:":"110.52.250.126","url:":"/static/image/common/logo.png"})
    client.put("ip_info", "3", {"ip:":"192.168.123.45","url:":"/uc_server/data/avatar/000/05/83/35_avatar_middle.jpg"})
```

⑦ 更新包索引，安装 Python 3 和 pip（如果 Python 3 安装后未自动安装 pip），操作命令如下。

```
sudo yum update
sudo yum install python3
sudo yum install python3-pip
```

⑧ 使用 Python 命令执行 text.py 文件，将本地数据导入 Hbase 中，操作命令如下。

```
python text.py
```

⑨ 切换到 HBase Shell 运行的终端窗口，执行 "scan 'ip_info'" 命令，查询 ip_info 表中插入的数据，如图 8-2 所示。

```
hbase(main):002:0> scan 'ip_info'
ROW                    COLUMN+CELL
 1                     column=ip:, timestamp=1678668957672, value=172.0.1.100
 1                     column=url:, timestamp=1678668957676, value=/static/image/common/search.png
 2                     column=ip:, timestamp=1678668957679, value=110.52.250.126
 2                     column=url:, timestamp=1678668957682, value=/static/image/common/logo.png
 3                     column=ip:, timestamp=1678668957684, value=192.168.123.45
 3                     column=url:, timestamp=1678668957687, value=/uc_server/data/avatar/000/05/83/35_
                       avatar_middle.jpg
3 row(s) in 0.0230 seconds
```

图 8-2　查询 ip_info 表中插入的数据

任务 8.4　流数据处理的简单应用

任务描述

使用 Flume、Kafka、Flink 进行简单的流数据处理，实现单词计数功能。

任务目标

学会使用 Flume 监听端口数据，存入 Kafka 主题，并使用 Flink 编程消费 Kafka 主题内容，进行简单的流数据处理，最后将结果存入 Kafka 的另一个主题中。

任务实施

微课 8-4 流数据处理的简单应用

在进行操作前,需确保 Hadoop、ZooKeeper 和 Kafka 集群已启动。

1. Telnet 工具的安装与使用

① 使用 "yum" 命令安装 Telnet 和 Netcat 工具,操作命令如下。

```
[root@master1 ~]# yum install -y telnet
[root@master1 ~]# yum install -y nc
```

② 启用一个本地的 TCP 端口 7777,等待客户端发起连接,测试 Netcat 工具能否使用,操作命令如下。

```
[root@master1 ~]# nc -lk 7777
```

③ 打开一个终端窗口,执行 "telnet 127.0.0.1 7777" 命令,进入 Telnet 客户端命令行模式,操作命令及结果如下。

```
[root@master1 ~]# telnet 127.0.0.1 7777
Trying 127.0.0.1...
Connected to 127.0.0.1.
Escape character is '^]'.
```

④ 在 Telnet 客户端命令行模式下向目标服务器发送信息,若要退出命令发送模式,则同时按 "Ctrl+]" 组合键即可,执行 "quit" 命令退出 Telnet 客户端,输入信息如下。

```
test
txt
```

若在 "nc" 监听窗口中监听到输出数据,则代表测试成功,操作命令及结果如下。

```
[root@master1 ~]# nc -lk 7777

test
txt
```

如果 Netcat 顺利监听到数据,则可终止上述命令完成测试。

2. Kafka 主题的创建与查看

① 创建 wordcount 主题,设置分区数为 4,副本数为 1,操作命令及结果如下。

```
[root@master1 bin]# ./kafka-topics.sh --create --zookeeper master1:2181,slave1:2181,slave2:2181 --partitions 4 --replication-factor 1 --topic wordcount
Created topic "wordcount"
```

② 查看已经创建好的主题列表,操作命令及结果如下。

```
[root@master1 bin]# ./kafka-topics.sh --list --zookeeper master1:2181,slave1:2181,slave2:2181
--consumer_offsets
test
wordcount
```

项目 8　Hadoop 平台应用综合案例

③ 查看 wordcount 主题的具体信息，操作命令及结果如下。

```
[root@master1 bin]# ./kafka-topics.sh --describe --zookeeper master1:2181,
slave1:2181,slave2:2181 --topic wordcount
Topic:wordcount	PartitionCount:4	ReplicationFactor:1	Configs:
Topic: wordcount	Partition: 0	Leader: 1	Replicas: 1	Isr: 1
Topic: wordcount	Partition: 1	Leader: 2	Replicas: 2	Isr: 2
Topic: wordcount	Partition: 2	Leader: 0	Replicas: 0	Isr: 0
Topic: wordcount	Partition: 3	Leader: 1	Replicas: 1	Isr: 1
```

3. Flume 的配置与启动

① 进入 flume/conf 目录，操作命令如下。

```
[root@master1 ~]# cd /usr/local/flume/conf/
```

② 新建文件 nc_wordcount.conf 并编辑，操作命令及编辑文件内容如下。

```
[root@master1 conf]# vi nc_wordcount.conf

a1.sources = s1
a1.channels = c1

a1.sources.s1.type = netcat
a1.sources.s1.bind = localhost
a1.sources.s1.port = 7777

a1.channels.c1.type = org.apache.flume.channel.kafka.KafkaChannel
a1.channels.c1.kafka.bootstrap.servers = master1:9092,slave1:9092,slave2:9092
a1.channels.c1.kafka.topic = wordcount
a1.channels.c1.parseAsFlumeEvent = false

a1.sources.s1.channels = c1
```

③ 启动 Flume（此时不要关闭操作窗口），并指定配置文件为 nc_wordcount.conf，操作命令及部分结果如下。

```
[root@master1 flume]# flume-ng agent -n a1 -c conf --f ./conf/nc_wordcount.conf
-Dflume.root.logger=info,console
......
2023-03-27 14:13:58,868 (lifecycleSupervisor-1-0) [INFO - org.apache.flume.
instrumentation.MonitoredCounterGroup.start(MonitoredCounterGroup.java:95)]
Component type: CHANNEL, name: c1 started
2023-03-27 14:13:58,869 (conf-file-poller-0) [INFO - org.apache.flume.node.
Application.startAllComponents(Application.java:182)] Starting Source s1
2023-03-27 14:13:58,872 (lifecycleSupervisor-1-1) [INFO - org.apache.flume.
source.NetcatSource.start(NetcatSource.java:155)] Source starting
2023-03-27 14:13:58,885 (lifecycleSupervisor-1-1) [INFO - org.apache.flume.
```

```
source.NetcatSource.start(NetcatSource.java:169)] Created serverSocket:sun.
nio.ch.ServerSocketChannelImpl[/127.0.0.1:7777]
```

4. 数据流写入 Kafka

① 打开一个终端窗口，执行"telnet 127.0.0.1 7777"命令连接对应端口，并向端口写入一些字符串类型的数据，操作命令及结果如下。

```
[root@master1 ~]# telnet 127.0.0.1 7777
Trying 127.0.0.1...
Connected to 127.0.0.1.
Escape character is '^]'.
hello world
OK
hello scala
OK
hello world
OK
hello world
OK
```

② 再次打开一个终端窗口，创建 Kafka 消费者，用于消费 wordcount 主题内容，操作命令及结果如下。

```
[root@master1 bin]# ./kafka-console-consumer.sh --bootstrap-server
master1:9092,slave2:9092,slave2:9092 --topic wordcount --from-beginning
hello world
hello scala
hello world
hello world
```

5. Flink 编程

① 在 IDEA 中创建一个 Maven 项目，将其命名为 FlinkWordCount，并在 IDEA 中安装 Scala 插件，勾选启用该插件。

② 在项目的 src/main 目录下新建一个目录，将其命名为 scala，选中该目录并右击，在弹出的快捷菜单中选择"Mark Directory as"选项。

③ 右击项目名，选择"Open Module Settings"→"Sources Root"选项，在进入的界面中选择"Modules"选项，单击"+"按钮，选择"Scala"选项，若本地没有 Scala，则单击"Create..."按钮，再单击"Download..."按钮，选择 Scala 的版本为 2.11.0，单击"OK"按钮。

④ 编辑 pom.xml 文件，编辑内容如下。

```
<?xml version="1.0" encoding="UTF-8"?>
<project xmlns="http://maven.apache.org/POM/4.0.0"
        xmlns:xsi="http://www.w3.org/2001/XMLSchema-instance"
        xsi:schemaLocation="http://maven.apache.org/POM/4.0.0
        http://maven.apache.org/xsd/maven-4.0.0.xsd">
```

```xml
        <modelVersion>4.0.0</modelVersion>
        <groupId>org.example</groupId>
        <artifactId>FlinkWordCount</artifactId>
        <version>1.0-SNAPSHOT</version>
        <properties>
            <maven.compiler.source>8</maven.compiler.source>
            <maven.compiler.target>8</maven.compiler.target>
            <project.build.sourceEncoding>UTF-8</project.build.sourceEncoding>
        </properties>
        <dependencies>
            <dependency>
                <groupId>org.apache.flink</groupId>
                <artifactId>flink-scala_2.11</artifactId>
                <version>1.10.0</version>
            </dependency>
            <dependency>
                <groupId>org.apache.flink</groupId>
                <artifactId>flink-streaming-scala_2.11</artifactId>
                <version>1.10.0</version>
            </dependency>
            <dependency>
                <groupId>org.apache.flink</groupId>
                <artifactId>flink-connector-kafka_2.11</artifactId>
                <version>1.10.0</version>
            </dependency>
        </dependencies>
</project>
```

⑤ 在 src/main/scala 目录下，新建 scala 文件，选择 Object 类型，将文件命名为 WordCount，并编写代码如下。

```scala
import org.apache.flink.api.common.serialization.SimpleStringSchema
import org.apache.flink.streaming.api.TimeCharacteristic
import org.apache.flink.streaming.api.scala._
import org.apache.flink.streaming.connectors.kafka.{FlinkKafkaConsumer, FlinkKafkaProducer}
import org.apache.kafka.clients.consumer.ConsumerConfig
import java.util.Properties

object WordCount {
  def main(args: Array[String]): Unit = {
    //创建流数据处理环境
    val env = StreamExecutionEnvironment.getExecutionEnvironment
    env.setParallelism(1)
```

```
    env.setStreamTimeCharacteristic(TimeCharacteristic.ProcessingTime)
  //配置 Kafka 连接属性
  val properties = new Properties()
  properties.setProperty(ConsumerConfig.BOOTSTRAP_SERVERS_CONFIG,
    "master1:9092,slave1:9092,slave2:9092")   //Kafka 集群
  properties.setProperty(ConsumerConfig.AUTO_OFFSET_RESET_CONFIG,
  "earliest")
  //从主题的第一条消息开始读取
  //配置键值反序列化
  properties.setProperty(ConsumerConfig.KEY_DESERIALIZER_CLASS_CONFIG,
    "org.apache.kafka.common.serialization.StringDeserializer")
  properties.setProperty(ConsumerConfig.VALUE_DESERIALIZER_CLASS_CONFIG,
    "org.apache.kafka.common.serialization.StringDeserializer")
  //创建 Kafka 消费者
  val consumer = new FlinkKafkaConsumer[String]("wordcount", new
      SimpleStringSchema(), properties)
  //消费 Kafka 主题 wordcount 中的内容
  val wc = env.addSource(consumer)
  //对 Kafka 主题 wordcount 的内容进行转换处理
  val firstProcess = wc.flatMap(data => data.split("\\s+")).map((_, 1))
  //按照单词进行分组
  val secondProcess = firstProcess.keyBy(_._1)
  //对分组数据进行聚合统计
  val result = secondProcess.sum(1).map(_.toString())
  //输出
  result.addSink(new FlinkKafkaProducer[String]("result_wordcount", new
      SimpleStringSchema(), properties))
  //执行任务
  env.execute()
  }
}
```

6. Flink 项目的打包、运行和结果查看

① 将提供的 JAR 包（配套资源：FlinkWordCount.jar）上传至容器内，并打开一个新的终端窗口，执行 "flink run -m yarn-cluster /opt/FlinkWordCount.jar" 命令，启动该任务，操作命令如下。

```
[root@master1 ~]# flink run -m yarn-cluster /opt/FlinkWordCount.jar
```

② 启动 Kafka，查看经过 Flink 流数据处理后的 "result_wordcount" 主题内容，操作命令及结果如下。

```
[root@master1 bin]# ./kafka-console-consumer.sh --topic result_wordcount
--bootstrap-server master1:9092,slave1:9092,slave2:9092 --from-beginning
(hello,1)
```

项目 8 Hadoop 平台应用综合案例

```
(world,1)
(hello,2)
(scala,1)
(hello,3)
(world,2)
(hello,4)
(world,3)
```

③ 再次打开一个终端窗口，执行 "telnet 127.0.0.1 7777" 命令，并在创建的端口中继续输入内容，操作命令及结果如下。

```
[root@master1 ~]# telnet 127.0.0.1 7777
Trying 127.0.0.1...
Connected to 127.0.0.1.
Escape character is '^]'.
hello world
OK
hello flink
OK
```

④ 查看 Kafka 的 result_wordcount 主题的结果。

```
(hello,1)
(world,1)
(hello,2)
(scala,1)
(hello,3)
(world,2)
(hello,4)
(world,3)
(hello,5)
(world,4)
(hello,6)
(flink,1)
```

至此，使用 Flume、Kafka、Flink 进行流数据处理的单词计数功能圆满实现。

项目小结

本项目介绍了 HDFS、Hive、MySQL、HBase 的数据互导，使用 Hive 进行了简单的数据分析操作，重点介绍了使用 Flume 监听端口数据，存入 Kafka 主题，再使用 Flink 编程消费 Kafka 主题内容，在进行简单流数据处理后将结果存入 Kafka 的另一个主题中的全部操作过程。学习完本项目，读者可为使用 Flume、Kafka、Flink 构建大数据实时处理系统奠定坚实基础。

课后练习

1. 请解释使用 Sqoop 将 MySQL 中的数据导入 Hive 时设置的如下参数的含义。

```
sqoop import
    --hive-import
    --connect
    --username
    --password
    --table
    --columns
    --m
    --hive-table
```

2. 请解释使用 Sqoop 将 HDFS 中的数据导出到 MySQL 时设置的如下参数的含义。

```
sqoop export #导出
    --connect
    --username
    --password
    --m
    --table
    --export-dir
```

3. 大数据处理的常用方法有离线处理和在线实时流数据处理两种。在互联网应用中，大数据的基本数据来源是日志数据（如用户的访问日志、用户的单击日志等）。请根据自己所学知识，利用 Flume、Kafka 和 Flink 构建一个大数据实时处理系统，并对设计思路做简要说明。